OLD GROWTH, NEW DIRECTIONS

OLD GROWTH, NEW DIRECTIONS

150 YEARS OF POPE & TALBOT

BY HARRY H. STEIN

Copyright ©2003 by Pope & Talbot, Inc.
All rights reserved. No portion of this book may be reproduced or utilized in any form, or by any electronic, mechanical, or other means without the prior written permission of the publisher.

Printed in China
09 08 07 06 05 04 03   6 5 4 3 2 1

Design: Patrick David Barber/pdbd
Cover photograph: Art Wolfe
Composition: Patrick David Barber/pdbd
Copy editor: Amy Smith Bell
Book title provided by: David Wright

Library of Congress Cataloging-in-Publication Data
Stein, Harry H.
      Old growth, new directions : 150 years of Pope & Talbot / Harry H. Stein.
         p. cm.
      Includes bibliographical references and index.
      ISBN 1-57061-386-9
      1. Pope & Talbot Lumber Co.—History. 2. Lumber trade—United States—History. 3. Lumber trade—Northwest, Pacific—History. I. Title: Pope and Talbot. II. Title.
      HD9759.P6S74 2003
      338.7'674'009795—dc21   2002191225

CONTENTS

FOREWORD BY PETER T. POPE	vii
SIGNIFICANT EVENTS IN POPE & TALBOT HISTORY	ix
INTRODUCTION	xv
CHAPTER ONE **FOUNDATION, 1849-1860**	1
CHAPTER TWO **DEEP CONNECTIONS, 1861-1880**	17
CHAPTER THREE **GROWTH, 1881-1924**	35
CHAPTER FOUR **TRAVAIL, 1925-1940**	61
CHAPTER FIVE **REVITALIZATION, 1941-1961**	83
CHAPTER SIX **UNCERTAINTY, 1962-1970**	105
CHAPTER SEVEN **SUCCESSION, 1971-1985**	117
CHAPTER EIGHT **NEW ERA, 1986 AND BEYOND**	145
EPILOGUE	161
DESCENDANTS OF DEACON PETER TALBOT AND COLONEL WILLIAM POPE	164
SELECTED BIBLIOGRAPHY	168
INDEX	169

FOREWORD

WHEN I CAME TO WORK FOR POPE & TALBOT IN 1960, THERE WERE MANY PUBLICLY traded forest-products companies that were family-owned or -controlled in the western United States. They included Bohemia; Brooks Scanlon; Crown Zellerbach; Dant & Russell; Pacific Lumber, Medford; and Willamette Industries. For various reasons all these firms have now disappeared. How is it that Pope & Talbot has survived when others in the industry have not? How did it manage to last through the financial panics of the nineteenth century, two world wars, the Great Depression, the postwar wave of mergers and acquisitions, the roller-coaster nature of the commodities industry, and the environmental challenges of the late twentieth century? How has the company lasted for more than 150 years? Why is it today the oldest forest-products company in the American West?

The immediate answer that comes to mind is luck—we have had our share! No company can survive for 150 years without some luck. We have also successfully maintained certain continuities that have supported each other throughout Pope & Talbot's history—continuities of principle, of family commitment, of management, of the willingness to change, and of perseverance. Let me say a brief word about each of these.

First—continuity of principle: Our New England heritage has given the company a determined but conservative direction. From the outset Pope & Talbot never equated size with success. My father, George Pope Jr., said many times that Pope & Talbot did not have to be the biggest—just the best! This lack of corporate ego, in my opinion, has kept the company from self-destructive gambles. We have maintained the traditional New England aversion to debt. This concept of carrying low debt has repeatedly saved the company during the many periods when prices paid for our products wreaked havoc among, even destroyed, our competitors. At the same time, Pope & Talbot labored to retain its reputation for corporate integrity.

Second—continuity of family commitment: Most family businesses eventually run out of capable family members who are interested in participating in the business. Pope & Talbot has not. And even among those family members not actively involved, stock ownership has not been taken lightly. Rather, it has been considered a privilege, an inheritance not to be squandered by one generation, but cherished and nurtured and passed on. Having Popes, Talbots, and Walkers—with their New England work ethic and long-term view—involved in the firm for 150 years has been critically important to the company's survival.

Third—continuity of management: Throughout its history Pope & Talbot has successfully promoted capable people from outside the controlling families into our top management. These notable individuals have included Edwin G. Ames, Hillman Lueddemann, Charles L. Wheeler, Ed Hunter, George H. Folquet, William A. Whelan, and Michael Flannery.

Fourth—the willingness to change: Pope & Talbot has known when to change with the times. From sail to steam to rail, from Maine to California, to Washington, to British Columbia, Pope & Talbot has known when to adjust to shifting conditions.

Fifth—perseverance: That is, taking the long view, and eschewing short-term gain in favor of long-term stability and continued existence. Neither the family nor outside leaders have allowed economic travail, family differences, or the pronouncements of Wall Street to bring down the ship.

As you read this book, I hope you can identify the continuities that have allowed this company to survive more than 150 years. Making it through the nineteenth-century winters of a small town in northern Maine was a good place to start.

—*Peter T. Pope*

SIGNIFICANT EVENTS IN POPE & TALBOT HISTORY

1849   Andrew J. Pope and Frederic Talbot arrive in San Francisco to test the lumber trade. They enter a ship lighterage business with Captain Josiah P. Keller and Lucius Sanborn.

1850   Frederic Talbot and Andrew J. Pope establish Pope & Talbot as a San Francisco lumberyard. With Captain William C. Talbot they enter the lumber and cargo trade. Frederic Talbot returns to New England.

1852   Andrew J. Pope marries Frederic and William C. Talbot's sister Emily. Pope & Talbot buys a schooner and recruits workers in Maine. With Josiah P. Keller and Charles Foster they establish the Puget Mill Company.

1853   Puget Mill Co. builds a sawmill and company town in what becomes Port Gamble, Washington Territory. Pope & Talbot acts as headquarters and handles the mill's and its own lumber sales.

1855   Talbot, Pope, Keller, and Foster reorganize as W. C. Talbot & Company in San Francisco for the lumber trade and ship chartering. Popes and Talbots invest heavily in California land and ranches and new California businesses.

1857   Puget Mill Co. opens a second sawmill in Port Gamble and expands its shipping fleet.

1858   W. C. Talbot & Co. temporarily operates a lumberyard in Victoria, British Columbia.

1861   Agents of Puget Mill Co. acquire the company's first Washington timberlands.

1862   Pope & Talbot is formed in San Francisco to replace W. C. Talbot & Co.

1865–66   William C. Talbot and Andrew J. Pope gain exclusive ownership of Pope & Talbot and Puget Mill Co. Resident manager Cyrus Walker obtains a tenth interest in Puget Mill Co.

1868–70   The original mill is dismantled and a larger sawmill opens in Port Gamble.

1874–80   Puget Mill Co. is reincorporated as the Puget Sound Commercial Company to handle the ships and steam tugs, and Puget Mill Co. to handle the mills and timberland. The latter buys and rebuilds a sawmill and secures timber at Utsalady, Washington Territory.

1878   Andrew J. Pope dies, putting William C. Talbot in control of the firms and allied businesses.

1881   On William C. Talbot's death, his son William H. Talbot begins a forty-nine-year domination of the firms and allied businesses, aided by his brother Frederick C. Talbot.

1884   Puget Mill Co. buys and upgrades a sawmill in Port Ludlow, Washington Territory, which it permanently closes in 1890.

1886   Puget Mill Co. and other mills on Puget Sound form a pool, the Pacific Pine Lumber Company. The Talbots acquire part-interest in rail-served California lumberyards.

1888–90   Pacific Pine Lumber Co. buys and expands the Grays Harbor Commercial Company in Cosmopolis, Washington Territory, which Puget Mill Co. executives and associates later control. Puget Mill Co. untangles cross-ownerships and turns divisions into separate affiliated companies with distinct responsibilities.

1891   Puget Mill Co. and other mills on Puget Sound become joint owners of noncompeting tugboats on the Sound.

1892   Puget Mill Co. stops buying Washington timberland. It pioneers in the South African market.

1895   Pope & Talbot and other firms replace the ineffective Pacific Pine Lumber Co. with another pool, the Central Lumber Company.

1901   The Pope & Talbot partnership becomes a corporation.

1906   Edwin G. Ames replaces the retiring Cyrus Walker as manager on the Sound.

1910   The company opens its own logging camps to supplement its use of contracted loggers.

1914–17   World War I boosts the company's output and profits while labor conflict before, during, and after the war embroils its operations. In 1917 the firm opens its first coordinated real estate development, Alderwood Manor.

1919   The Port Gamble and Port Ludlow mills start lightering lumber on rail car barges to Seattle.

1920–24   Intermittent negotiations proceed to sell Puget Mill Co. and other properties. A Seattle sales office opens. The firm opens a Seattle residential development, Broadmoor.

1925   The Charles R. McCormick Lumber Company of Delaware buys both mills and about half of the company's timberland. William H. Talbot and George A. Pope Sr. become major stockholders in McCormick Lumber.

1926–29   McCormick replaces the mill in Port Gamble, modernizes Port Ludlow's mill, and expands its own Saint Helens, Oregon, mill. It expands and integrates McCormick's logging and shipping operations and greatly increases cutting and output. On William H. Talbot's death in 1929, George A. Pope Sr. assumes the presidency of the families' remaining firms.

1930-33   Pope, Talbot, and Walker interests assert control over their former properties and over McCormick's separate shipping firms.

1934-37   McCormick Lumber Co. permanently closes the Port Ludlow mill, shuts down all its logging camps and the San Francisco lumberyard, and takes other steps to survive.

1938   Pope & Talbot Lumber Company is established to take over the bankrupt McCormick Lumber Co. A court awards it and the McCormick steamship companies to the unpaid Pope, Talbot, and Walker interests.

1940   Pope & Talbot, Inc., is incorporated to replace McCormick Lumber Co. Puget Mill Co. is dissolved, and McCormick Shipping continues separately under the control of the Popes, the Talbots, and the Walkers. George A. Pope Jr. becomes Pope & Talbot president, and George A. Pope Sr., until his death in 1942, board chairperson.

1941-45   The now prospering firm operates seventy-six ships, runs busy mills and timber cutting, and at war's end opens a tree farm around Port Gamble. Talbot Walker and then Charles L. Wheeler lead the company during George A. Pope Jr.'s absence in the army.

1946-47   Pope & Talbot booms during the postwar era. The company begins a ten-year, $26 million piecemeal expansion, modification, replacement, and diversification program. It follows the trees into Oregon, acquiring the Penn Tract near Oakridge. It re-enters coastal, intercoastal, and Latin American shipping routes, acquires war-built ships, and resumes the sale of logged-over Washington land.

1948-50   A sawmill and, successively, a particleboard plant and green veneer mill are built at Oakridge, and a big tree farm is opened on the nearby Penn Tract.

1952-57   The company abandons all but one maritime route and sells many vessels. Using the ship proceeds plus a stock buyback, Pope & Talbot partially liquidates.

1960   At Oakridge the company becomes the nation's first manufacturer of medium density fiberboard for the furniture industry.

1961   The company enters plywood manufacturing by buying and upgrading a plant in Kalama, Washington. It closes its last shipping line and leases and then sells all remaining vessels.

1963   Corporate headquarters moves from San Francisco to Portland. Cyrus T. Walker becomes president, and George A. Pope Jr., chairperson of the board. The company liquidates an affiliated fertilizer company.

1967   A subsidiary Pope & Talbot Development is established to energize property development schemes. It opens the first phase of Port Ludlow, a residential and recreational community.

1969   In its largest acquisition in 120 years, the company buys and upgrades sawmills, with attendant cutting rights, in British Columbia at Midway and Grand Forks, and also buys and quickly closes a lumber company at Westbridge, B.C.

1971   Peter T. Pope becomes chairperson and chief executive officer, and Guy Pope, president and chief operating officer when Cyrus T. Walker and George A. Pope Jr. retire. The Saint Helens sawmill and nearby forestland are sold.

1972   The company buys a stud mill in Hudson, Ontario.

1974   The Hudson stud mill is sold.

1978   The company begins a ten-year capital expansion plan. It buys half-interest in a pulp mill in Halsey, Oregon. George A. Pope Jr. dies. Guy Pope leaves management, and William A. Whelan replaces him as president.

1979   Pope & Talbot reincorporates in Delaware. It closes the Kalama plywood mill.

1980   The company buys and upgrades paper mills in Eau Claire and Ladysmith, Wisconsin, to make private-label absorbent products. It buys and rebuilds a sawmill in Spearfish, South Dakota.

1983   The company buys the remaining half of the Halsey pulp mill.

1984   R. Steven Mason replaces William A. Whelan as president and chief operating officer.

1985   Pope & Talbot spins off the tree farms, the Port Gamble mill site and town, Port Ludlow, and the company's undeveloped acreage to the new Pope Resources. It leases the Port Gamble mill and undertakes management of the historic town.

1987   Pope & Talbot is listed as a Fortune 500 corporation.

1988   The company buys a paper-tissue mill in Ransom, Pennsylvania. It opens a disposable diaper plant in Shenandoah, Georgia, and buys similar plants in Aiken, South Carolina; Maryville, Missouri; Oneonta, New York; and Porterville, California.

1989   Pope & Talbot sells the Oakridge mill and the South Carolina diaper plant. It buys a sawmill in Newcastle, Wyoming.

1992   The company buys a sawmill, with attendant cutting rights, in Castlegar, British Columbia.

1993   The company sells the last of the Penn Tract acreage.

1995 The company closes the Port Gamble sawmill. Michael Flannery replaces Peter T. Pope as president. Pope remains chairperson and chief executive officer.

1996 Pope & Talbot sells all of its diaper plants.

1998 The company buys majority ownership of Harmac Pacific, owner of a giant pulp mill in Nanaimo, British Columbia. It sells the two Wisconsin tissue mills and the Ransom, Pennsylvania mill.

1999 The company sells the Halsey pulp mill in a sale, lease, and buyback arrangement. It buys the remaining ownership of the Harmac Pacific mill. Peter T. Pope retires from the company, and Michael Flannery becomes chief executive officer.

2000 The company closes the sawmill in Newcastle, Wyoming.

2001 Michael Flannery becomes chairperson of the board.

INTRODUCTION

BRITISH COLUMBIA, CALIFORNIA, OREGON, AND WASHINGTON POSSESS SOME OF THE MOST spectacular coniferous forests on earth. From Northern California far into Canada, an immense softwood forest blankets the region between the Pacific Ocean and the Cascade Mountains. Trees carpet shoreland flats, valleys, peaks, and islands. A relatively mild temperature and western maritime winds generate heavy rain. Climate and soil together encourage year-round growth of fir, hemlock, cedar, and spruce. This temperate zone contains a dynamic landscape of forests containing both old and new growth. Trees of a dominant species die or are cut and other species replace them. For centuries Indians, wildfire, windstorms, insects, disease, and grazing animals created a patterned succession of forest vegetation and structure.

A range of factors determine the exact composition of a particular stand. Each stand typically mixes growing and full-grown trees with standing dead trees (snags), thick understories, and, rotting on the ground or in streambeds, fallen trees and limbs. In some stands big trees overwhelm smaller ones. Trees can become too closely packed for any to mature to great size. Some species are more tolerant of shade than others and displace competitors. Fires choke off underbrush and clear out young trees to benefit larger and older ones. The wildland fire suppression of the twentieth century created dense and stagnant stands as well as heavy falls of insect-damaged trees.

Old-growth forests during the nineteenth-century settlement era contained huge trees, many centuries old. Port Orford cedar and Douglas fir were already well established in the time of Columbus. Western red cedar and coast redwood were already leaving mulch during the early Middle Ages.

Pope & Talbot ship in rough seas.

The big Northwest old-growth trees always dwarfed the loggers.

xvi INTRODUCTION

The abundance, height, girth, and straightness of these trees amazed early settlers and loggers. Four-to six-hundred-year-old Douglas firs averaged 200 feet in height—some soared to 300 feet—and were 5 to 12 feet in diameter. Tall firs were free of limbs for at least their lower hundred feet. Four-hundred-year-old hemlocks were 200 feet tall. Red cedars rivaled California's giant redwoods. The coast redwoods of California and Southern Oregon were free of limbs up to 150 feet and more above the forest floor. Giant sequoia, or Sierra redwood, had average diameters of a5 to 20 feet. Settlers actually built cabins in the redwood hollows.

Indians made incidental cuts in these great forests. Beginning in the late 1700s, American, English, Russian, and Spanish explorers and traders made tiny inroads along the Pacific coastline. In 1850, when lumbering began in earnest, old-growth forest covered 15 to 24 million acres, some 60 to 90 percent of the Pacific Northwest. Western Washington and Oregon contained 80 percent of the nation's Douglas fir, the most important of the western species. Douglas fir was the world's greatest lumber tree, ideal for heavy and light building, spars, masts, and other uses. Initially, hemlock, cedar, and spruce were not considered commercially useful except to make shingles. Wood was the universal building material of the era: cheap, easy to work, and abundant.

The discovery of gold in California in 1848 forever altered these old-growth forests. The Gold Rush affected communities and peoples on a national and international scale. It created an unprecedented demand for building and mining materials in California, initially at exorbitant prices. Eastern lumber transported around Cape Horn to San Francisco in mid-1849 sold at prices second only to gold. At this time San Francisco lumber merchant Andrew J. Pope questioned the need to bring shipments around Cape Horn when substantial resources were so conveniently located for exploitation. One western redwood tree yielded up to 300,000 board feet, double or triple the cargo a ship could transport from Maine. The vast Oregon County, Pope assured associates in the East, "will supply this State with all but pine B[oar]ds at as low [a] rate as they can be got here from the States." On a scouting expedition to Puget Sound to find and occupy a heavily forested lumber mill site, Pope & Talbot partner Captain Josiah P. Keller was pleased with what he saw. "We shall make some handsome lumber here," he said in 1854, "if the rains do not prevent us." The steam mills of George Meigs, Andrew J. Pope, William Renton, William Sayward, Asa Mead Simpson, William C. Talbot, and Henry Yesler would drive the commercial lumbering and settlement of what would become the Washington Territory.

CHAPTER ONE

FOUNDATION

1849-1860

ORIGINS

POPE & TALBOT'S MAJOR FOUNDERS CAME FROM FAMILIES OF SEA CAPTAINS, MERCHANTS, sawmill operators, and ship owners on the Machias River in coastal Maine. In the seventeenth century the first Peter Talbot and Ralph Pope arrived in the mainland colonies from England. Colonel Frederick Pope and Captain Peter Talbot, their grandsons, commanded troops at the Battle of Lexington and fought throughout the American Revolution. On the Machias, Popes joined with Talbots, and they with the neighboring Keller, Foster, and Chaloner families, in businesses, marriage, and town affairs. All these families would help found the Puget Mill Company and Pope & Talbot, man their vessels, and locate, build, and work their mill in Port Gamble, Washington Territory.

Tidy, conservative villages like East Falls and Lubec spotted the thinly settled pine region of Maine. Close-knit, intermarried families tended to dominate Maine's villages. Its early sawmill towns were steeped in personal and family connections, paternalistic stewardship, and class gradations. Communities were meant to be disciplined, moral, productive, harmonious, and profitable places.

Many villagers on the Machias River farmed part-time and worked seasonally on shares in the woods and in their own small sawmills and tidewater shipyards. This was the era of the ax- and saw-wielder, of winter ox teams yarding logs, of spring river drives, sorting booms, and bluewater log rafting. Later, these logging techniques would transfer to Puget Sound.

The Talbots supplied wood products to late colonial New England and New York. In the young republic their vessels entered the Caribbean. Maine's renowned lumber, lime, ice, and granite reached far-flung ports. Bangor, Bath, Calais, Ellsworth, Machias, and Portland thrived as shipping points by the 1840s. Many Pope and Talbot boys, after minimal schooling, first saw the larger world as common seamen on family ships.

Entrepreneurial-minded clans such as the Talbots and the Popes powered Maine's economic takeoff during the 1820s. In Massachusetts and Maine economic rank and reputations for careful morals and prudent business practices aided the Talbots and the Popes to become church elders, militia leaders, council members, legislators, and prominent lawyers, doctors, and officials. Before 1854 the Talbots and the Popes were usually Whigs, occasionally Democrats, and one was an antislavery Free Soiler. By the Civil War they and their sons in California were fervent Unionists and Republicans.

The Talbot family's boatyard occupies the foreground, and the Pope family's sawmill lies at midstream, in this painting of East Machias, Maine, in 1855. *(East Machias Public Library)*

Business nevertheless came first. In 1849 Andrew J. Pope and Frederic Talbot, ambitious and fully supported by their families, set out for California's Gold Rush. They went west as bearers of such traditional New England values as industry, prudence, restraint, truthfulness, and thrift. Both Pope and Talbot accepted the ways of early, small-scale New England capitalism. They had its aptitude for risk-taking, profit-seeking, and entrepreneurship. They adopted its simple partnership arrangements and reliance on family capital. The migrants served as New England's agents in exploiting markets, natural resources, and the maritime trade.

BEGINNINGS

ANDREW J. POPE AND FREDERIC TALBOT'S JOURNEY FROM THEIR NEW ENGLAND HOMES was arduous. The men took ocean passage to Central America, crossed Panama, and on the Pacific side boarded ship for San Francisco. California was wild with the promise of gold. Within eighteen months of the initial discovery, eighty thousand people flooded into the Gold Rush country. By August 1849 San Francisco had mushroomed from a sleepy village into a ramshackle port of six thousand residents. By December the population had passed the twenty-thousand mark.

Ten thousand New Englanders joined in the Gold Rush. A few hundred of them were not gold seekers but mill owners looking for new markets. By 1849 Maine was nearing the end of its heyday as a shipping and manufacturing center and just beginning to decline from its historical status as the nation's premier lumber producer. Yankee mill owners realized that the tiny water-powered mills of the Far West could not satisfy California's building and mining needs. In 1849 Bangor mill owners sent thirty-nine ships, with 5 million board feet, around the Horn to California. A typical cargo included clapboards, framing lumber, planed boards, joists, window frames, doors, slashes, nails, grindstones, tobacco, stoves, and liquor.

Pope and Talbot were older than most of the young men making their way to California at this time. They intended to be sojourners, not permanent settlers, and to return home with a fortune. Sharing profits, these capable individuals expected to build markets and revenue for their families in the East. Although they brought little capital with them—Pope carried only a few thousand dollars—they were better prepared than many of their competitors. They had reliable credit and supply sources, access to salable cargoes, sturdy ships, and loyal crews.

An inadequate number of docks on San Francisco Bay presented an initial difficulty—and opportunity. Anchoring in the overcrowded harbor, ships had to unload onto rafts and small boats. On December 3, 1849, Pope, Talbot, and Maine neighbor Lucius Sanborn entered this lighterage trade. For offloading purposes, the partners bought longboats, scows, and yawls. After splitting $800 in profits two months later, Talbot and Pope bought out Sanborn, purchased additional boats and a small river sloop, and began selling consigned products.

More than eight hundred ships, many abandoned, occupy Yerba Buena Cove along San Francisco's sprawling waterfront in 1851. Warehouses, hotels, and prisons fill some abandoned vessels. *(San Francisco Maritime National Historical Park)*

Tents cover beaches around San Francisco in 1851, when the city's swelling population put building at a premium. The scene was a lumber merchant's dream.

They also helped fill the huge demand for wood products. A thousand tents covered the San Francisco dunes. Jury-rigged homes, saloons, theaters, and gambling dens could not be thrown up fast enough. Large fires repeatedly leveled these structures. Wooden sidewalks had to wait until lumber was less costly. Booming mining districts could not wait. On January 4, 1850, twenty-nine-year-old Andrew J. Pope and thirty-one-year-old Frederic Talbot established the first enterprise to be named Pope & Talbot. Even at his young age, Pope was already a veteran businessman. Early on, this handsome, dark-haired six-footer had chosen to follow his brothers into his father's business. Unlike many of the Pope and Talbot men, the sea was not for him.

On a beach lot the partners handled lightering and sold lumber and other items from Maine and as far away as New Zealand. A single cargo cleared between the low hundreds and the low thousands of dollars. Their sloop took lumber upriver to Sacramento and Stockton, then the urban entryways to the mining districts. The Pope & Talbot partnership also invested in dirt-cheap city lots and in undeveloped acreage around San Francisco Bay.

Thirty-four-year-old Captain William C. Talbot, Frederic's brother, brought the brig *Oriental* into the bay in March 1850 and used his stake in the vessel to buy into the firm. That autumn he pioneered Pacific coast lumber sales to the Kingdom of Hawai'i. When Frederic Talbot departed in December to open a commission house in New York City, William Talbot took his place. By his thirties William Talbot was a veteran captain and a respected lumber trader on the East Coast, in Central America, and in Europe. Of medium height, he had dark, curly hair, dark eyes, a high forehead, and a powerful build. Years at sea had given him a ruddy complexion and a commanding presence. Ashore, he spoke quietly.

> *"I have often speculated on what made my great-grandfather, Andrew J. Pope, and my great-great-grandfather, Captain William C. Talbot, leave the comfort of small-town life in Maine. I think three factors influenced their decision. First, they were the younger sons in their families, and it was the older sons who had the good jobs. (Primogeniture was still practiced in New England.) Second, there was tremendous hype in the eastern press at that time about California. Third, the two men could see that Maine was running short of timber, and there had been reports of great stands of timber in the Northwest. But even with all that, at the time they left, Andrew Pope and William Talbot were not making a decision to move permanently; that decision would come later.*
>
> *I have visited East Machias, Maine, and I can tell you they made the right decision. The population of East Machias is smaller today than when they left it more than 150 years ago. The only employment is picking blueberries."*
>
> —Peter T. Pope

Selling wood products on consignment and on its own account, Pope & Talbot thrived in a rough-and-tumble trade. In 1852 it left the beach and opened a brick warehouse and lumberyard at Pine and Battery Streets. San Francisco's population increased that year to forty thousand. On November 19, Pope and Talbot joined with Josiah P. Keller and Charles Foster, of the East Machias, Maine, Foster & Keller shipyard, to form the Puget Mill Company "for the purpose of manufacturing Lumber in the Territory of Oregon at Puget Sound." California's non-Indian population now numbered about a quarter-million people. The potential customer base was both large and obvious. Other lumber merchants also initiated backward vertical integration of their holdings by constructing mills and exploiting timber in the Pacific Northwest.

News reports of the timber on Puget Sound had reached Maine much earlier in the nineteenth century, however, without arousing notable attention. Andrew Pope's enthusiasm changed that thinking at home. Why import lumber from New England when it could be produced at less cost on the Sound, he contended. Talbot family relative Lafayette Balch already had a profitable piling trade and other interests on the Sound when Frederic Talbot and Pope rented his San Francisco cottage in December 1849. Pope and Talbot were used to seeing rough lumber cargoes arrive in California from the Northwest well before they made their decision to found the Puget Mill Company.

Little capital was needed to open a ship-dependent "cargo" mill. The Puget Mill Company was capitalized at a mere $30,000. Invested funds went to acquire machinery, hire personnel, and construct a small mill and a few buildings in Washington Territory. Pope & Talbot held $20,000 of the stock; Foster & Keller the remainder, acquired largely via signing over their yard's newly completed schooner, the *L. P. Foster*. The sheltered land at Teekalet Bay, originally a federal land grant held in Keller's name, cost nothing.

In 1853 Pope & Talbot dispatched Captain William C. Talbot and the schooner *Julius Pringle* to find and occupy a mill site on the Sound. The *Foster*, meanwhile, took on mill machinery and workers in Maine. Keller sailed to join Talbot as a partner and mill manager. In June 1853 the *Pringle* scouted among the bays and inlets of the upper Sound and along the Hood Canal, the latter a sixty-two-mile-long natural channel separating the Olympic and Kitsap Peninsulas. The Maine natives aboard saw broad sandy beaches, numerous rivers, and on the western horizon the soaring Olympic Mountains. Salmon filled the waters. Solitary seals, ravens, and eagles broke the silence. Several hundred Americans, at most, resided on Puget Sound. The expedition passed a few natural harbors with good shelter and thick adjacent forest. Claimants had preceded Talbot to Port Townsend, Port Ludlow, and Seabeck.

The size and lushness of the forest extending down to the beach and covering the headlands staggered the newcomers. Fir and hemlock covered the higher ground; alder, maple, and cedar, the valleys and lowlands. The rich soil produced a luxuriant, nearly impenetrable mass of vines, shrubs, and young trees. Rhododendron, salmonberry, huckleberry, Oregon grape, bracken, swordfern, and blackberry thrived among willow, hemlock, and cedar.

On the Kitsap Peninsula the *Pringle* party settled on a large spit of sandy land beside sheltered Teekalet, or Gamble, Bay. The bay had sufficient room and depth—five feet on average and nine feet in some parts—for docking vessels in a protected anchorage. The location, barely five miles from the mouth of the Hood Canal, had the advantage of shortening voyages for sailing ships previously bound to inlets farther south. Pope & Talbot called the place Teekalet, "brightness of the noonday sun" in the local S'Klallam language, until officially changing the name to Port Gamble in 1868.

By the time the *Foster* arrived, Keller found the pioneer *Pringle* party hard at work on the mill site. From the *Foster*, common laborers and skilled workers joined millwright E. S. Brown, machinist James White, future manager-partner Cyrus Walker, and others of the *Pringle* group. Keller unloaded a complete steam-powered sawmill, supplies, and store merchandise. The captain settled in as resident manager with his wife and daughter—the first of Teekalet's non-Indian women—and young son.

Just above tidewater, Keller and Talbot erected crude quarters, a cookhouse, a general store, and the sawmill itself, a 45- by 70-foot structure. Two dozen men, women, and children constituted Teekalet's original population. When local Indians resisted removal to their treaty-established reservations, Keller gave them lumber to build homes on land across Gamble Bay. Paddling across from the Little Boston village, S'Kallams kept the mill running when white employees deserted during the 1858 Fraser River Gold Rush. By necessity, Keller and his successors also hired African-Americans, Sandwich Islanders, and Chinese immigrants for rough mill and dock labor and mess hall and hotel chores.

In its first year, 1854, the mill ran for eleven months, averaging 15,000 board feet a day. (One board foot equals a one-inch-thick, one-foot-long board, twelve inches in width.) Except for what was sold to settlers or used to expand the town, everything went by ship to domestic and foreign markets. In 1854 Port Gamble loaded 3.6 million board feet of lumber, 42,103 board feet of piling, 64,000 shingles, and 223 masts and spars. The next year Pope & Talbot built a second and substantially larger mill at Teekalet.

Pope & Talbot reorganized in 1855 as W. C. Talbot & Company, with Foster and Keller as junior partners. (The Pope & Talbot name was officially restored in 1862.) It moved into a Stewart Street warehouse-lumberyard among other dealers. Talbot & Co. also entered into ship chartering and supply-purchasing activities. Rice, candles, ham, bacon, pork, beef, flour, and sugar passed through the warehouse. While Talbot & Co. apparently did well, the Puget Mill Company was definitely a profit-maker from the start. The mill was "paying very well now," Andrew Pope noted in 1856, "and if she does not burn up she is bound to pay well." Captain William Talbot quit the sea and joined Pope in the office. Although they visited Port Gamble on rare occasions, both men preferred the comforts of San Francisco over life in that isolated and rainy place.

The transplanted New Englanders helped make San Francisco the economic hub of the Pacific coast. Like western mine owners, Puget Sound manufacturers attached heavy spokes to the city by the bay. Regional lumber production increased from 26 million board feet in 1849 to 239 million board feet in 1859. Although impressive to westerners, the performance still represented a negligible portion of the decade's yield in the eastern states.

Washington became, and remained, a resource hinterland tributary to San Francisco. For many decades the city was the economic capital of the Pacific Slope, commanding regional trade, commerce, and finance. Talbot and Pope were among those who dominated the industry on Puget Sound. "Later augmented by purchase of plants at Port Ludlow and Utsalady," forest historian Robert E. Ficken writes, the Port Gamble mill "would be the leading force in that California-dominated industry for the remainder of the nineteenth century."

EXPORTS

TALBOT AND POPE EXPECTED SAN FRANCISCO TO BE THEIR PREMIER MARKET, BUT THEY soon learned not to totally depend on California. Without foreign customers far western sawmills were at the mercy of a volatile coastal market. The sawmills were located in a region isolated from the eastern states and subject to frequent and dizzying economic swings. Regional conditions, especially in California, counted more than national trends, at least until the 1880s, when the Pacific Northwest was at last integrated into the national economy. The Panic of 1857, seriously rocking the settled states, still made times in the West "rather blue," according to Andrew Pope.

Despite this, between 1856 and 1861, the co-managers increased production at Port Gamble and obtained more vessels for their lumber fleet. The Nevada silver-mining boom restored Gold Rush conditions to California. In 1864 alone a thousand new buildings went up in San Francisco. The state experienced as many downturns as upswings, however, in the years before and during the Civil War.

Western cargo firms like the Puget Mill Company inevitably looked offshore to supplement their San Francisco sales. Their location allowed them to exploit Pacific Rim markets closed to manufacturers on the Atlantic coast. Puget Sound producers turned to foreign markets whenever California demand dropped. When demand was high in San Francisco, manufacturers, with one major exception, slighted such places.

Port Gamble sent 40 percent of its lumber abroad in the mill's first year of operation. Australia was "all the rage" due to the discovery and development of gold deposits. The first Puget Mill cargo was off-loaded in Sydney, after sixty days at sea, in September 1854. "We shipped 2000 ft[.] to Manilla [sic]," Pope reported early in the new year, adding that he expected to be able to "market considerable Lumber in China" in the near future.

The California Gold Rush peaked that year. Prices fell and banks failed. By February 1855, Pope advised, times were "rather tight." The only thing to do was "to hold on and Weather the Storm." The firm invariably took the long view: Develop business connections and keep foreign demand supplied, even if no immediate profit resulted. Pope reasoned that overseas markets would eventually be vital. Even a meager rate of exports kept Port Gamble running and its labor force employed. Pope & Talbot plowed profits back into the company, snapped up cheap California land, and added domestic and foreign trade capacity. Talbot arranged long-term connections with agents in Australia, China, and Hawai'i—Hackfeld & Company was the main affiliated firm in Honolulu.

The Puget Mill Company led the way in penetrating Pacific Rim markets. In 1855 it dispatched six cargoes to China, clearing $16,000 on the voyage of the *Live Yankee* alone. An unrelenting competitor, it made enemies, especially on the China coast. "Look out for those d____d thieves," William J. Adams of the Washington Mill Company warned. Claiming that the Puget Mill Company had deviously undercut him in China, Adams vowed to "get even as sure as there is a God in Israel."

Port Gamble in 1855 furnished masts and spars to England's famous screw-paddle wheeler the *Great Eastern*, the world's largest-tonnage vessel. Chile, reported Pope, paid "some profitt [*sic*] when freight is but 10$ . . . and logs at 5$." The mill sold lumber in Tahiti and Australia and in 1857 sent another hundred masts to England. Of the 7.9 million board feet cut in 1857, 5.5 million board feet went abroad. Port Gamble provided more than half the lumber exported from the West Coast in that year.

Hawai'i was, year by year, the best foreign market, even when competition was intense. Pope claimed that the company realized more money per board foot in the islands than anywhere else on the Pacific Rim. As late as 1920 the Washington lumberman J. H. Bloedel asserted that Hawai'i was a perpetual gold mine for his historically venerable rival.

CAPITALISTS

SMALL PARTNERSHIPS WERE POPULAR AMONG EARLY NEW ENGLAND CAPITALISTS. SUCH relationships made sense in villages ordered by complex personal and familial bonds and for enterprises needing modest financing. The western partnerships were also family-based, with interest shared between the two coasts and effective control resting in the San Francisco Pope & Talbot headquarters. The partners trusted one another enough to maintain simple business arrangements. Occasional profit-and-loss accountings were entirely satisfactory, as were rudimentary documents listing individual shares in the accumulating properties.

Marriage strengthened the smoothly functioning Talbot-Pope relationship. On a visit home in 1851, Andrew Pope married William Talbot's sister Emily. The brothers-in-law built San Francisco mansions beside one another. Their partnership with Keller and especially with Charles Foster, however, underwent strain. Keller and Foster, moreover, had differences of their own. At Port Gamble, Keller, pressured by personal creditors to wring more money from the mill, was frustrated by Pope's seeming placement of the lumberyard's interest above that of the mill's. In 1854 he complained to Foster that "Pope carries on his business entirely independent of the P Mill Co. & buys of them what lumber he wants [while charging a high wharfage for its ships at his San Francisco docks]." Pope retorted that he had, after all, invested the most capital in the concern. After five years on the Sound, Keller returned to Maine to look after affairs at his shipyard. For his part Foster grew increasingly distrustful.

The fact that the Puget Mill Company basically functioned as a Pope & Talbot subsidiary accounted for much of the strain. The San Francisco firm controlled sales, finances, shipping, land, logging, and milling activities in both California and Washington Territory. The mill cut timber and lumber only when Pope & Talbot sent orders and installed only the machinery dictated by San Francisco. Because of slow

From left to right:
Andrew J. Pope (1820–78)
Captain William C. Talbot (1816–81)
Cyrus Walker (1827–1913)

communications with New England, Talbot and Pope made key decisions without consulting Foster.

Talbot and Pope honored their respective family traditions by placing long-term gain over immediate returns, whether in profits, payout, or administration. Debt was to be avoided at all costs. "Times are rather tight here and anyone has to have the real back bone to get along these times," Andrew Pope said in 1855. "I do not see much chance to make Money but the Main Thing is to hold on and Weather the Storm, which I think I shall be able to do as I have no Bank Notes out, *and do not want any.*"

Not wanting their sons to become rich drones, the fathers brought them into the office to learn the business as prospective successors. They wanted family men to hold authority in both Port Gamble and San Francisco. No relative should come from New England "with large ideas about wages," Pope cautioned. "Dead heads" were not wanted, either on the Sound or in California.

The senior partners became rich and respected members of San Francisco society. The local newspapers treated the Popes and the Talbots as community leaders, moving assuredly among wealthy bankers, lawyers, and businesspeople. The society pages featured their spacious mansions, fashionable clubs, ranches, and seashore retreats. Talbots and Popes married into other well-known California families, including that of railroad developer Henry Mayo Newhall. No Talbot or Pope ever considered residence at Port Gamble.

The San Francisco *Morning Call* estimated Andrew Pope's worth at between $1 and $4 million in 1871. He therefore held second ranking among California millionaires, behind the famous Big Five of transcontinental railroad fame. He and Talbot owned the mill, lumberyard, commission business, numerous ships, and stock in various major corporations. Even if the lumbering venture had ultimately failed, the wild appreciation in the value of their urban and rural land holdings alone would have made them very wealthy.

A specific property's potential, rather than its current value, guided buying decisions. Pope and Talbot shrewdly anticipated San Francisco growth patterns and invested heavily when prices fell during periodic economic downswings. They scooped up acreage at tax and foreclosure sales and moved the lumberyard when the land became more valuable for other purposes. Holdings ranged from platted lots and undeveloped tracts

Captain William C. Talbot's San Francisco mansion, pictured here late in the nineteenth century. Andrew J. Pope lived nearby, in another palatial mansion.

in San Francisco and the East Bay to ranchland in the coastal and San Joaquin valleys. In the mid-1870s Pope owned land valued at $668,500. Talbot's investments were in the same range. With late-coming mill partner Cyrus Walker, Pope and Talbot jointly owned additional real estate worth more than $720,000.

A preference for tangible properties such as sawmills, ships, and real estate made the partners invest in a far more conservative fashion than was the norm among wealthy Californians. By and large, they avoided mining stocks. Pope was an incorporator of the state's first insurance company in 1861, the California Mutual Marine Insurance Company and, three years later, the state's first commercial bank, the Bank of California. He also held stock in the California Dry Dock, the Merchant's Exchange, the Pacific Silk Manufacturing Company, and the Central Rail Road Company.

PORT GAMBLE

BUILDING A TOWN AND FURNISHING ESSENTIAL SERVICES WERE MANDATORY ENDEAVORS on isolated Gamble Bay. Nearly every structure in Port Gamble occupied company land, secured at no cost by Keller under the Oregon Donation Land Act. Only employees or workers in necessary businesses and services were allowed to reside in the town. The employees lived in assigned company housing, dined at the company mess hall, purchased provisions at the company store, played on the company baseball team, and traveled on company vessels. Life revolved around the mill's routine. A whistle blast woke the village at 5:20 A.M., six days a week. Twenty minutes later a second whistle sent single men and transients hurrying to the mess hall for breakfast. Work started precisely at 6 o'clock, the beginning of an eleven-and-a-half-hour day. Whistles signaled the half-hour lunch break and, late in the day, quitting time. Community celebrations were limited to Christmas, Thanksgiving, and Independence Day.

In 1857 the company built a second, bigger mill farther out on the spit, adding workers and housing. The village contained 42 buildings and 202 residents, slightly more than half of whom were either born in or with parents from Maine. A warehouse, carpenter shop, cookhouse, blacksmith shop, and commercial flour plant clustered about the mill. By 1862 the Puget Mill Company employed 102 men in the mills and woods. Single men lived in crude cabins or a dormitory. Cottages and more substantial dwellings, some with solid New England furnishings, appeared on the bluff above the original Teekalet site.

An idealized Port Gamble in 1856, as the town is cleared.

PORT GAMBLE 11

Single men had to eat in Port Gamble's mess hall, seen in 1908. Food was plain, abundant, and quickly eaten. Until the 1930s Chinese men and women cooked and served the meals.

To attract families, the firm organized Kitsap County's first school in 1862. Other amenities included a barber shop, a bowling alley, a daguerreotype studio, and a dance hall. "The Boys are very fond of dancing," Andrew Pope explained in relation to the latter innovation, "and we have to go to considerable expense to keep them quiet." Port Gamble also featured boardwalks, a community hall for worship and other public functions, and picket fences to control the company pigs and horses. Young maples shaded the streets and flowers graced the yards.

For decades a pall of smoke, ash, and cinders wafted into the village and along the shore. In certain winds and seasons charred sawdust fell on the town. To prevent fires in ships, woods, and mills, Port Gamble could do little except sweep beams and floors, dispose of waste, and guard candles and lanterns. "It is a perfect terror here on a windy night," observed Cyrus Walker, especially "when the sawdust is blowing a living sheet of fire" from the burner.

By 1870 most of the original dwellings had been replaced by permanent wooden structures. Garbage-loving pigs still jostled people in the cookhouse alley, but Port Gamble had two private hotels and a Masonic lodge. The town boasted of its school, Congregational church, library, theater, and brass band. In addition to selling staples, the store carried books, Maine and New York newspapers, and the latest fashions. By this time barely a third of the town's 246 residents claimed Maine origins. The non–New Englanders included thirteen Chinese men and twenty-nine Hawaiians.

The narrow front-gabled houses, white clapboard facings and picket fences, maple-lined streets, and enduring Maine connections caused visitors to think of Port Gamble as a transplanted New England village. In truth, Port Gamble more closely resembled a sawmill settlement than a typical Yankee town with a central village green and a church and solid commercial establishments. No village green existed in Port Gamble. The wharf, mill, and housing for unmarried men occupied the spit, the closest thing to a scenic site. Residents looking down from the bluff associated the spit with roistering seamen, stevedores and loggers, drinking and gambling in a private hotel saloon.

Employees in Washington mill towns probably began playing baseball soon after the Civil War. Port Gamble's baseball club, a member of the Sawdust League, is pictured here in 1894.

The company ideal, however, stressed an attractive, disciplined, moral, and stable Port Gamble, reflecting the Calvinism of the founders. Management assumed that orderly conditions would help attract and retain responsible workers and protect the village from undesirable influences. To Talbot, Pope, and especially Cyrus Walker, Keller's successor as manager, work was a blessing, not a curse. The longer an individual toiled, the less time the worker had to squander on bad or wasteful activities. The mill owners considered themselves stewards of employee welfare, exercising a benevolent paternalism. Those disliking the arrangement were invited to draw their pay and leave.

In the same fashion as at "home" in New England, good workers secured promotions. The firm encouraged single men to marry, form families, and eventually send their boys to work at the mill. Workers hurt in accidents or enfeebled in loyal service were given undemanding employment. Management discouraged misbehavior but conceded the impossibility of stamping out all bad habits. Adulterers and prostitutes were obliged to depart. Keller and Walker restricted gambling and drinking. By hiring and paying off their ship crews in Port Townsend, they avoided formation of a sailors' district in Port Gamble. But the managers dared not prohibit liquor.

The company built many simple worker cottages such as these in Port Gamble. (John D. Cress, photographer)

Government, whether national, territorial, or local, exercised little influence. "We must see that we are not required to do too much," Keller reflected, because "we are to be the taxpayers in our County." Dedicated to low taxes, the mills controlled Kitsap County for years. Until the 1860s, when logging on public land was first challenged, the federal government was not an issue.

Only once, after hundreds of Puget Sound Indians rebelled against the making of treaties by Governor Isaac Stevens, did mill executives actually solicit help. For two days in the fall of 1855, whites took shelter in a makeshift Port Gamble blockhouse, awaiting an attack that never came. Management usually trusted the local S'Klallams but still feared that raiders would burn the mill in the night. As a means of encouraging the war effort against the hostile Indians, the Puget Mill Company sold substantial amounts of supplies, on credit but at high cost, to both the territorial volunteers and the regular troops. Although the affair had nothing to do with the local conflict, which had ended the previous spring, Haidas from British Columbia and sailors from the U.S.S. *Massachusetts* fought an amphibious engagement on Port Gamble Bay in November 1856. The townspeople buried Gustavus Engelbrecht, the lone American casualty of the battle.

Neither Keller nor Walker intended the town to be a duchy. Walker ruled more by example than by decree. Yet when he put his hands in his pockets and spit, employees watched out. Even in his seventies, Walker intimidated people. On labor-short Puget Sound, workers often quit because of unfair or harsh treatment. Traditions of independence ran deep in the mills. Ties of blood and affection among the immigrants from Maine undermined most attempts at arbitrary control by the bosses.

Managers, supervisors, and skilled workers, such as sawyers and filers, occupied two-story dwellings like these in Port Gamble. *(John D. Cress, photographer)*

The company preferred to hire employees from Maine, or at least from England, Scotland, and Wales. The workers actually available on Puget Sound were of "rather poor quality," Andrew Pope reported. Natives of Maine, in contrast, embodied the virtues of dedicated labor, loyalty, and harmonious living. Great effort went into securing sawyers, filers, master mechanics, blacksmiths, and ax and boom men, all traditional Yankee occupations. Men, and some women too, "were constantly . . . urged to come out and work for the firm." Employees visiting Maine, a former executive recalled, were instructed to "tell anyone familiar with milling or logging to come along and they would always be taken care of." Travel loans and other inducements were available. Jacksons, Thompsons, and Walkers followed one another to Port Gamble. For three generations migrants came from a Maine undergoing serious decline from sustained overcutting in its pineries and sawmills.

"Clan Machias," detractors claimed, held the best jobs in the mill and on the ships and tugs. The mess hall catered to their taste for "baked beans, johnnie cake, and cod fish." Maine folks dominated the clubs and other social institutions. Truly a clannish lot, they little tolerated marrying anyone not from the Pine Tree State. One story told of a man falling into the mill pond, splashing about and shouting that he was drowning. Nobody paid much attention until someone remembered his pedigree and yelled, "Gosh, Almighty! He's from Machias! Sound the whistle!"

CHAPTER TWO

DEEP CONNECTIONS

1861-1880

WAR AND SEPARATION

THE SAN FRANCISCO FIRMS AMPLY REWARDED THEIR NEW ENGLAND FAMILIAL CONNECtions. The Pope mills in Maine consigned hardwood lumber, doors, and windows for them to sell. Payments and drafts cleared through East Coast–based Pope and Sons. Talbot trading firms supplied food, stoves, turpentine, candles, and nails for sale at Port Gamble. The Popes, Talbots, Fosters, and Kellers built and owned ships in common that served their interests on both coasts.

The easterners counted heavily on West Coast remittances within months of the mill's startup. Andrew J. Pope conceded in 1855 that the monthly $15,000 sent over the past year had become vital to the family's well-being. Although he cautioned them against involvement in risky ventures, Pope's relatives bought a money-losing mill in Nova Scotia, suffered financially during the Panic of 1857, and by the end of 1860 faced actual ruin. The eastern Talbots also needed bolstering from time to time. Pacific Rim earnings were crucial to both families after the outbreak of the Civil War in 1861 disrupted East Coast shipping patterns.

For Talbot and Pope the Civil War presented both problems and opportunities. Pope assured eastern relatives that he would sell all his California real estate in the unlikely event the bottom dropped out of the family's eastern firms. But failure of the East Machias or Boston business, in which he held a minority interest, might easily force liquidation in the West, he worried.

Talbot and Pope increased production to meet California's expanding demand. The state's economy had returned to the heights of the original Gold Rush years, and by 1860 San Francisco was home to more than fifty thousand people, making it one of the nation's largest cities. Port Gamble shipments reached 19 million board feet in 1862. The two men also bought land, thousands of acres along the California coast and in the San Joaquin Valley. Between additional purchases and the appreciation of existing holdings, the value of Pope's holdings alone grew from $50,000 in 1861 to $118,000 three years later.

Josiah P. Keller struggled to keep Port Gamble operating during the Civil War. "Why should all business stop," he asked, "because war prevails?" Higher wage demands and the loss of workers to mining rushes perturbed him. "The mines! the mines!! & higher wages is the watchword." Urged on by Keller, Port Gamble's Maine natives contributed heavily to the U. S. Sanitary Commission, an organization formed to aid wounded Union troops. The village actually claimed to have given more per capita to this cause than any

San Francisco lumber merchants cluster along Stewart Street in 1867. W. C. Talbot & Company had been at 149 Stewart since 1856.
(T. E. Hecht, photographer; San Francisco History Center, San Francisco Public Library)

other urban locale in the country. In San Francisco, Republican brothers-in-law Pope and Talbot had taken prominent places in Lincoln's 1860 campaign for California's electoral votes. Both participated actively in Unionist affairs: electing Northern sympathizers, organizing the legislature, and establishing armed volunteer units to head off secessionist conspirators.

Three deaths in 1862 suddenly and radically changed matters. The demise of Keller; of Samuel Pope, an East Machias mainstay; and of Pope's father, William, automatically dissolved all the partnerships. For more than two years Talbot and Pope negotiated a settlement with several groups of heirs. They bought all outstanding shares of the Puget Mill Company and of a renamed Pope & Talbot. Though a debt-hater, Pope pledged to pay the easterners $320,000, with the final payment due in 1872. In 1865 the heirs accepted his cash, promissory notes, and shares in the eastern businesses. The shrunken Atlantic coast assets, including Pope's holdings, totaled $538,227. The western assets, excluding real estate belonging solely to Captain William C. Talbot, were inventoried at $443,559—a staggering sum for a decade of work.

In 1863 Talbot had coaxed Cyrus Walker into overcoming his own dread of debt and accepting a $30,000 loan to buy a tenth of the Puget Mill Company. A successful acting manager between 1858 and 1861, Walker was Keller's obvious successor at Gamble. Talbot and Pope wanted the mill to be run by a person with a vested interest in the concern. They held the conventional wisdom that a man with his own business was more of a man than one satisfied to be an employee. A new joint agreement in 1864 apportioned Talbot and Pope 45 percent each in the Puget Mill Company and 50 percent each in Pope & Talbot. The former agreed to pay a 5 percent commission to Walker on all transactions.

To Washingtonians, Walker *was* the Puget Mill Company for half a century. Born in Madison, Maine, in 1827, he moved restlessly about the country in his early years. He worked on farms and in sawmills and as a logger, schoolteacher, surveyor, and starch factory manager. Arriving aboard the *Pringle* in 1853, Walker helped construct the mill and served as timekeeper, accountant, and general handyman. The personification of the New England work ethic, he pushed everybody hard, himself the hardest. Walker originally wanted to earn $50,000 on Puget Sound and then return to Maine. In 1858 he began speculating in land in three-year-old Seattle. Between real estate and the flourishing sawmill, he made a fortune.

The war years and family circumstances forever changed matters. By 1865 Pope & Talbot and the Puget Mill Company were entirely western-owned. Their emancipation from financial ties to the eastern families represented one of the most significant changes in a century and a half of corporate history.

GETTING THE LAND

THE PUGET MILL COMPANY OPERATED IN A WASHINGTON TERRITORY WHERE TIMBER wealth was all but free for the taking. With the exception of Native Americans, northwesterners generally considered frontier land a public commons, free for everybody to use if it was unsurveyed and unreserved. People helped themselves to the timber around Puget Sound in much the same fashion that miners, farmers, and ranchers helped themselves to other resources in the public domain.

The first Puget Sound mills obtained no land beyond their immediate locales. Their limited capital went into buildings, equipment, and ships. If the land was unclaimed, or claimed but unoccupied, people simply took the trees. If settlers had claimed the land, the mills bought cheap logs or stumpage from them. Removing timber from unprotected acres became a tradition and an alleged right. Under one pretext or

The *H. D. Bendixsen* loads for San Francisco at Port Gamble in 1890. Lumber went into its stern and over the deck rail. Loading or discharging lumber from any sailing vessel normally required several weeks.

THE KING PHILIP: A SHIP'S TALE

The 694-ton schooner *Kitsap*, built for Pope & Talbot in Port Ludlow in 1887, sails on the Puget Sound between 1910 and 1915. She carried more than a million board feet to South America, Australia, and Asia. *(San Francisco Maritime National Historical Park)*

POPE & TALBOT SHARED IN THE SOMETIMES DARK lore of America's sailing era. The 1,241-ton *King Philip*, a full-rigger built in Maine in 1856, sailed unnoticed until the crew set her afire in Honolulu in 1867. The company purchased her cheaply in 1869 and repaired and modified her for the non-wood trade.

The *King Philip* was put into the West's burgeoning wheat trade and to carry general merchandise on its second or return legs. Bound for Liverpool in 1870, she endured five months of heavy weather and stayed afloat only by jettisoning 250 tons of wheat. For the next few years she sailed quietly between America's two coasts. Or she carried guano from Howland Island to Hamburg and returned to the East Coast to load coal or oakum and pitch for San Francisco's shipbuilders.

On May 16, 1874, the *King Philip*, according to Frederick C. Matthews's 1931 account, "sailed from Baltimore with a riotous crew, being followed down the bay by boarding house runners in boats" probably seeking money the seamen owed. On May 18, while the ship lay in Chesapeake Bay, crew members set fire to the cargo. Officers and sailors extinguished it. Ordered to weigh anchor, all but three seamen refused, saying that she was too badly damaged to proceed to sea. Because such disobedience was typically treated as mutiny, U. S. Marines came aboard. "After considerable parlaying the crew was overawed," and the alleged chief conspirator was put ashore. The rest of the crew wanted to quit but were kept aboard under guard, as the captain reported that he wished no further expenses. Matthews did not mention when the Marines disembarked.

Onward to California, heavy gales stunned the vessel off Cape Horn. Many spars and sails disappeared overboard. Five feet of water settled into the well. The exhausted ship put about for distant Rio de Janeiro and four months' repair. The *King Philip* finally arrived in San Francisco in May 1875, 351 days—not the usual two months—from Baltimore.

Diverted to the calmer West Coast lumber trade, the *King Philip* came to an end in a notorious marine graveyard. Captained by A. W. Keller, nephew of company founder Josiah P. Keller, she was being towed in ballast from San Francisco harbor on January 25, 1878. Released from the tug's hawser just inside the Golden Gate, Keller had her barely under way when the wind completely died. She lost way and dropped two anchors, but heavy seas parted the chains. She drifted solidly onto a sandy shore near the Cliff House. The breakers quickly wrecked her. The ship remains sold for $1,500. Keller was given a new command.

A tallyman counts lumber removed by longshore workers from the company's *Okanogan*. The four-masted schooner, launched in 1895, is unloading at Pope & Talbot's giant San Francisco yard at Third and Berry Streets. *(San Francisco Maritime National Historical Park)*

another, timber was readily and freely obtained. An 1831 federal law prohibiting removal of timber from the public domain was not enforced. Trespassing was so common as to be recognized in regional folklore. Most residents of western Washington believed that much good and little harm came from harvesting timber in the seemingly inexhaustible forest. The great trees, like hoped-for land grants and railroads, were the means of realizing a grand destiny of population growth and wealth.

Obtaining legal title to timberland in antebellum times was generally impossible and, in any event, unnecessary. Much of the land in Washington was unavailable for legal sale because the dense and tangled undergrowth made the required preliminary surveys difficult. Furthermore, federal law favored agricultural, as opposed to industrial, exploitation. Keller referred to Teekaleet, which was a part of the Donation Land Act, as "our farm." No federal law provided specifically for timberland sale until 1873. But there were many ways to obtain logs.

By the Civil War many of the mills wanted to secure land to guarantee their supply of raw material. No firm was more determined than the Puget Mill Company. By 1875 it had become the largest Washington timber-holder. In a territory wanting to sell land to build schools, a university, and other projects, mills found plenty of willing sellers. The Civil War also stimulated the opening of federal land as a means of generating revenue. Washington Territory offered 46,000 acres for sale in 1861; the proceeds were dedicated to funding a territorial university in Seattle. Buyers selected unlimited amounts, paying $1.50 an acre. The Puget Mill Company acquired 15,260 acres in Kitsap and nearby Island, Jefferson, and Mason Counties. Meanwhile, 3 million acres of federal land in the territory went on the market between 1863 and 1871. The four biggest Puget Sound mills were the largest purchasers, Puget Mill topping the list with 74,693 acres at $1.25 an acre.

Settlers, loggers, and lumber companies still logged the "commons," resisting occasional attempts to enforce the laws against trespassing. Charging criminal theft, federal officials in 1861 secured the indictment of Keller, Pope, and other prominent mill owners. The prosecution, however, was not pressed, since juries in

Loggers needed relentless strength, and aging loggers had to keep up or find other work. Two loggers from the company's Camp Talbot in 1928, having undercut the tree, use a crosscut saw to fell it in traditional fashion. *(Walter P. Miller, photographer)*

logging counties were unlikely to convict anybody of timber theft. Puget Sound loggers and mill operators eventually pled guilty to federal theft charges in return for nominal punishment. Walker recalled that several of the miscreants were quickly released after smoking cigars with the judge and the sheriff in a locked Port Townsend cell. Revenue-hungry wartime officials substituted an honor code, selling public logs in roundabout fashion. Until 1865 Washington residents paid the federal treasury fifteen cents per thousand board feet. The Puget Mill secured more than 2 million board feet for $325 in depreciated greenbacks.

The postwar West remained true to the national tradition of legally and illegally exploiting the land laws. Washington lumber firms could no more conceal their land machinations than their output of finished product. Widespread violation of the law was an open secret. Federal investigators charged in the late 1870s that $40 million worth of timber had been stolen from the public domain on Puget Sound, with the Puget Mill Company the principal culprit.

In Washington, as elsewhere in the West, firms used employees or hired outsiders to secure free land under the Homestead Act of 1862 and to buy acreage under the Timber Culture Act of 1873, the Timber and Stone Act of 1878, and other statutes. Dummy entrymen received from $50 to $125, plus expenses, to register for and obtain patents. Third-party agents then received the vital documents for eventual transfer to the ultimate holders. The Port Gamble mill boldly sent people to the land office. A steamer sailing from Seattle in November 1881 took, according to a local newspaper report, "a crowd of men from the Puget Mill Company up to Olympia on land office business."

Almost half of the 186,000 acres owned by the firm in 1890 had been acquired, at $2.50 per acre, under the Timber and Stone Act, legislation supposedly designed to prevent such acquisitions by industry. Anticipating the law's eventual repeal, Walker urged his associates to make "hay while the sun shines." Timberlands were booming, he warned in one of many letters to San Francisco on the subject, and Portland timber sharps and Great Lakes interests were especially active. The federal government twice suspended Timber and Stone Act entries. In 1888 the company made a hefty contribution to the Republican presidential candidate Benjamin Harrison, correctly anticipating in return a lifting of the suspensions on its filings. The firm also commenced purchases of privately held acreage, obtaining that year a large block of land along the Chehalis River in southwestern Washington. One way or another, it meant to get the timber.

STRUGGLE FOR ORDER

EVEN AS WASHINGTON'S MILLS EAGERLY ACQUIRED TIMBER, PRODUCTION BARELY increased between 1869 and 1879, from 129 million board feet to 160 million board feet. The number of operating sawmills declined, and only one major new plant opened. Downturns in the California market hurt, and the transcontinental railroad arrived too slowly for the opening of business to the Midwest. Americans, to be sure, wanted what sawmills made, as per capita consumption of commercial wood products rose by a factor of four in the second half of the nineteenth century. Yet domestic demand could not generate sustained good times in the face of geographical and transportation impediments.

"The most common experience of businessmen in the lumber industry," Washington historian Norman H. Clark concludes, "was ruinous competition, then failure, not success." Markets and prices gyrated. Lumber output often outpaced demand. Determining that exports could not absorb overproduction, operators decided that they must find ways of coping with excess output.

Americans of the Gilded Age were unaware of the fact that they lived in an era of long-term deflation. A prolonged depression after the Panic of 1873 ushered in a national trend of declining prices in goods and services. This circumstance continued, with intermittent periods of recovery, until near the end of the century. The depression of the 1870s dulled the Northwest lumber market, keeping prices low. The real estate boom of 1884 to 1885 in southern California briefly helped the mills, but the economic swings worsened thereafter, as transcontinental railroads began full integration of the Northwest into the national economy. The region was, to an extent greater than ever before, exposed to the full impact of national depressions.

In the 1870s larger lumbering firms swallowed smaller ones or expanded on their own accord. Companies cut prices when necessary, charged what customers could bear when possible, and generally added new functions and markets. "Whenever the market turned downward," historian Thomas R. Cox writes,

In 1876 the Puget Mill Company acquired a bankrupt mill on Camano Island and big timber stands nearby on the Skagit River. Reopened in 1879 with improved equipment and facilities, the Utsalady Mill lasted only another decade before closing in 1890. *(MSCUA, University of Washington Libraries, UW18511)*

"running time had to be reduced, profits dropped, and the less stable mills had to struggle to stave off bankruptcy."

Responding to their industry's fragmented, cutthroat, and profligate ways, mill owners formed pools modeled on those of the railroads, steel producers, and whiskey distillers. Pope & Talbot and the Puget Mill Company figured prominently in a succession of combinations, all legal at the time. Pools restricted output and, if successful, raised prices. Puget Sound sawmills joined with San Francisco dealers in 1871 to form the first pool, agreeing to ship and sell no more lumber in the city than the market demanded. The results were unimpressive. Adding additional Washington mills to the combine, San Francisco business interests tried again in 1877, only to have this latest effort fail within a year's time.

In 1880 the Pacific Pine Manufacturers Association pooled eleven mills in a better-regulated combine. Two years later *West Shore* magazine reported that association mills controlled prices for "the sugar pine of the Sierras, the redwood of the Humboldt district and the fir of Puget Sound." The collapse of the southern California real estate boom terminated the arrangement. In 1886 the Puget Mill Company tried again. This time they brought thirteen Washington and Oregon mills together to form the Pacific Pine Lumber Company. This concern attempted to control shipments to, and prices in, the key San Francisco market and to allocate foreign orders among the members. Many of the participants chose to dump poorer, sap-filled grades and otherwise unwanted stock on the combine, ignoring the solemn promises made in Pacific Pine meetings.

Each of the "gentlemen's agreements," the Pacific Pine Lumber Company excepted, collapsed within a year or two. Pools gained favor during depressed periods and lost appeal when prosperity returned. The many who refused to join, and the members willing to violate agreements, hampered attempts to effectively restrain trade. Although a handful of San Francisco–owned cargo mills on Puget Sound and on Humboldt Bay in northern California dominated the coast, they lacked sufficient concentrated power to truly dictate production and prices.

WESTERN WASHINGTON

A Douglas fir log, 10 feet in diameter, on the Port Gamble headrig in 1878. Thick blades guaranteed mounds of sawdust around all the saws and on employees.

"THE LOGS ARE NOW DRAWN UP A SLIP FROM THE *water into the mill upon an endless chain armed with grappling dogs, and as they one by one fall upon the bed are seized by steel teeth of live rolls, which at the motion of brakes toss backward and forward, like playthings the largest [log], leaving them just right for the cross cut [sic] saws which governed by brakes dart upward from their hiding places and in a jiffy the log is ready for the rotaries. From the rotary [saws] the lumber, save square timber, passes to the gang edgers, drawn along by live rolls. Beyond the edgers are stationed the markers, one to each saw, [workers] who cull and mark the lumber, from whom it is hurried along the live rolls to the sluices, down which it slides to the wharf, where one or two vessels are constantly loading. Not the least interesting features of these mills are their admirable arrangement of endless chains for carrying off the bark, top, but[t]s, slabs, edgings, sawdust, etc."*

—from *The Oregonian*, January 5, 1880

This depiction of a typical 1880 Puget Sound sawmill would have applied to some of No. 2 Port Gamble, the firm's largest mill. The depiction also typified once-bankrupt mills that the Puget Mill Company acquired and modernized in Port Ludlow and Utsalady, Washington Territory.

By 1880, Northwestern mills avoided eastern mills' machinery, which lacked the power and strength to handle their huge logs. As in pioneer days, No. 2 heaved logs out of log ponds into range of the headrig's principal saw. It cut logs into cants, large slabs with one or more rounded edges destined for other saws. Big steel hooks ("dogs") held the log securely on the steam-powered carriage serving the saw.

No. 2 used the double circular headrig common in large western mills. One unprotected saw blade sat above the other, rotating in the opposite direction. These rotaries cut huge logs that had been impossible for the original up-down "muley" headrigs to handle. Some competitors added a third one. A pony gang ("live gang") saw cut smaller logs, thus bypassing the headrig. Gang saws, two or more saws mounted together, cut logs and cants.

Master mechanic William Walker, Cyrus Walker's brother, designed the edgers for No. 2. Edgers squared a cant's edges and ripped it into lumber. A scantling machine made studs for framing buildings, ship framing, and the thicker posts and beams exported to Australia. Trim saws lowered to cut lumber or slabs into various lengths. Planers surfaced rough lumber. Another device machined tongues and grooves into boards. Machines also produced countless wedges for tree felling and millions of thin laths for wall plaster backing and pickets for fencing.

Sluices carried everything down to No. 2's dock. At Port Ludlow and Utsalady elevated railroads serviced the docks. Everything but timbers and flooring tumbled there into "random piles" for laborious hand-sorting by grade and length. Roughly sawn products were shipped wet and green because the three mills had neither sheds nor kilns for drying and because wood cargoes did not pay by weight. Nearly all the bark, edgings, sawdust, and other refuse went into a burner operating night and day.

Yet working in No. 2 was much like working in earlier mills. In exchange for steady wages, hard, bruising effort was common and taken for granted, at least by managers. Laborers removed loose bark manually, as machines and men turned the logs and cants. With little more than small slings, horse-drawn wagons, and muscle-power, longshoremen or sailors required days to load a ship. People handled wood at every step.

Equipment lacked modern safeguards. For endless hours "doggers" rode the sometimes bucking log carriages inches from the fearsome saw. Back and hand injuries and hearing losses were common. Boilers could explode. Operating or splintering saws, hundreds of unguarded belts, and tumbling and splintering logs or wood could cripple or kill even the wariest. Inexperienced, hungover, and playful workers added to the hazards.

Almost everybody endured the cold and wind penetrating through logways. Dampness came from the bay and from steam. Sawdust covered people and surfaces. Hour after hour, saws shrieked and screamed, belts thundered, and headrigs shook the floors. In fall and winter work began and ended in darkness. Supervisors pushed production and added night shifts; output rose impressively at No. 2. Yet nobody could command a steady pace; "cut to order" and "hurry up and wait" prevailed before the 1890s. Ship arrivals and departures largely determined all three mills' cycles. No storage facilities existed to regularize work flows. Thirty-six hours typically passed from receiving an order to felling trees to loading the results.

Sawmill technology altered quite gradually during the nineteenth century. It changed slower still on a Puget Sound reached erratically by engineering publications. Furthermore, Puget Mill Company executives and employees preferred to study and copy up-and-running mills. They relied on buying and upgrading equipment from bankrupt mills and on their own inventiveness. Typical in American manufacturing, the mechanically minded among the employees regularly tinkered with and improved devices. Cyrus Walker viewed labor-saving devices cautiously. Costly new alterations like band saws were introduced slowly, and only after years of consideration. The machine and blacksmith shops busily repaired, rebuilt, and refurbished equipment to keep operations running.

During the 1880s competitors in Washington installed band saws. They ran even faster, demanded less power, and, by creating far less sawdust than predecessors, increased salable lumber. Teeth cut in the long, looped, steel-bladed band saw could cut any cant and the biggest logs available. (Saw blades had insertable teeth to allow sharpening and easy replacement.) Walker still held fast to a forty-year company value, and urged president William H. Talbot in 1892 to avoid the expensive band saws being installed by competitors by the 1880s unless a new mill was built.

By 1899 company mills could fill orders for 304 dimensions of lumber, some in more than one grade or length. They resisted increasing their planing capacity, partly relying on planing facilities in Pope & Talbot's San Francisco lumberyard to add value to their shipments. The real benefit occurred

Another view of the big logs. *(John D. Cress, photographer)*

in Chicago, New York, Saint Paul, and other distant cities, where northwestern wood products became furniture, cabinets, and other highly manufactured items.

Competitive demands forced the Puget Mill Company to finally add specialized products. Bit by bit, it revised familiar techniques and installed higher-level equipment. Port Gamble and Port Ludlow early in the new century built sheds to air-dry green lumber. Some time before the First World War they installed kilns, which seasoned lumber by heating and withdrawing moist air. They then added molders to make trim to hide joints and use as decoration. They began mechanizing handling. And before that war they finally adopted band saws.

The Puget Mill Company itself increased output during the era of the pools. Between 1869 and 1884 it built, purchased, and expanded mills on Puget Sound, at the same time adding oceangoing vessels and tugs. In the winter of 1870 the No. 2 Port Gamble mill supplanted the pioneer 1853 plant, doubling capacity. Important acquisitions followed: in 1876 at Utsalady on Camano Island to the northeast, and in 1878 at Port Ludlow on the Olympic Peninsula. Both mills came cheaply. The firm bought the former Utsalady spar manufacturer, along with large timber stands on the Skagit River, for $32,000. Utsalady reopened with a new headrig for large-sized logs and other improvements in equipment and supporting facilities. Talbot prevailed on a reluctant Pope to pay $64,850 for the idle Ludlow mill. Outfitted with new machinery, the plant and company town resumed operation in October 1883.

Before Port Ludlow came on line, the firm produced barely 100 million board feet of lumber per year. By the mid-1880s, however, Gamble alone turned out 160,000 board feet a day. Ludlow produced 100,000 board feet and Utsalady nearly 75,000. In total, the company produced each year more than Greece or Portugal consumed annually. The overall yearly output actually declined, though, in large part because of cutbacks mandated by depressed national economic conditions. Without storage facilities none of the mills had space for the accumulation of inventory.

SUCCESSION

ABOARD THE GILDED AGE'S ECONOMIC ROLLER COASTER, MANY OF THE SMALLER MILLS adopted cut-and-run tactics, depleted timber, and went bankrupt. The Utsalady and Port Ludlow mills had fallen into the company's hands through bankruptcies. Shrewdly anticipating long-term increases in timberland values, Talbot, Pope, and Walker carefully husbanded the firm's acreage, especially those tracts that had been purchased at cheap rates. The mills instead mainly used logs bought on the open market.

The founders were determined to conserve their timberlands, run their mills well, and pass their properties on to the next generation. Talbot and Pope, the "lumber kings of the Pacific Coast," according to one San Francisco business publication, favored the long view. They frugally controlled capital and operating costs, avoided borrowing, declined to cut their own timber, and trained successors. The "business was rolling up like a great big snow ball," Pope often remarked to Walker, and his "own anxiety was as to who would run it some day."

William H. Talbot (1858–1930) dominated company affairs for forty-nine years.

The death of Pope or Talbot would have automatically forced the reorganization or sale of their empire. The aging partners contemplated their mortality—and incorporated. The corporate form was now firmly established as an instrument of the American economy. Compared with a partnership, the corporation limited individual liability and risk. A corporation closely held by their families would not threaten the ownership and a family leadership succession.

Steam rises from the Port Gamble mill in 1907 as a small ferry nears docked sailing ships. "Little Boston," in the foreground, houses the mill's S'Klallam workers and their families.

Two California corporations were established, retaining the old 45–45–10 percent ratio between Pope, Talbot, and the indispensable Walker. The Puget Mill Company was incorporated in 1874, and the Puget Sound Commercial Company in 1877, each for $2 million. The former company got the mills and timberland and the latter the sailing ships and steam tugs.

Soon afterward, in 1878, Andrew J. Pope died at the age of fifty-eight. Apparently by prior agreement, Talbot paid $40,000 for Pope's half-interest in the lumberyard. And he took William H. Talbot, at eighteen years his eldest son, into the San Francisco office. Young Talbot thereby took precedence for the eventual succession. Fourteen-year-old George A. Pope, Pope's oldest male child and heir, would have to wait.

The elder Talbot died at sixty-five in 1881. Under the captain's will, half of his Pope & Talbot holding went to son William H. and half to nephew Charles F. A. Talbot, already a company executive. The latter's share would, after ten years, automatically transfer to Frederick C. Talbot, William's younger brother. In 1885 fifty-eight-year-old Cyrus Walker married William's thirty-seven-year-old sister Emily. Their only child, Talbot C. Walker, was born the next year. Thus Andrew Pope's widow, Emily Talbot Pope, whose son George had just entered the business, became Cyrus Walker's aunt.

SUCCESSION 29

LOGGING IN NINETEENTH-CENTURY WASHINGTON

"BALKED BY THE TALL TIMBER OF GREAT GIRTH . . . *early loggers thinned out smaller trees with familiar tools and methods. River drives were impracticable, and logs were skidded over crude roads designed for the purpose. These skid roads, formed of short logs half-imbedded in earth and greased with the oil from dogfish liver, made it possible for 10 or 12 oxen to drag even the largest logs to the waterways.*

'Fallers' chopped the trees down, while 'buckers' cut or 'bucked' them into 24-, 32-, or even 40-foot lengths, using a crosscut saw, or 'Swede Fiddle.' In the late sixties, some weary axman borrowed a saw and discovered an easier and more efficient method of felling. From that time, fallers worked on springboard scaffolds, set high off the ground to save clearing underbrush and to avoid the pitchiness of the lower tree trunk; this type of cutting not uncommonly left stumps 15 feet in height.

The eighties marked the initial use of power equipment in logging. The earliest logging locomotive was built at Marysville in 1883. . . . The donkey engine appeared at Bellingham Bay in 1887, and soon, on Grays Harbor and lower Puget Sound, adaptations followed. First tried was the Dolbeer donkey, an upright steam engine with a capstan. Improvements brought greater power, substituting a windlass, and the engine was mounted on skids for mobility. A 'choker' loop of cable encircling a log was fastened to the long line, which the engine coiled, dragging in the log and 'yarding' it.

High-lead logging was begun in 1896 near Port Townsend. Instead of drawing logs with a straight, level pull, the cable from the donkey-engine drum was reeved through a huge block suspended from a spar tree and thence to the log. An even faster method, using the 'skyline,' by which a slingload of logs can be lifted and carried bodily on an aerial train, clearing underbrush, gullies, and streams, was introduced later."

—from Writer's Program, Works Projects Administration, *The New Washington: A Guide to the Evergreen State* (Hillsboro, Oregon: Binford & Mort, 1950)

Certain features fixed the pattern of logging on nineteenth-century Puget Sound: the methods brought from elsewhere; the geography and climate; the unique qualities of the trees; the employers' demands and facilities; and the local traditions. Adjacent timber camps might use different methods, while camps distant in place and time might employ the same ones.

Contractors largely replaced settlers as the Puget Mill Company's suppliers. The first loggers were clannish veterans of New England forests; later ones were experienced Midwestern- and Scandinavian-born loggers. Men from the Machias, Maine, area who logged for Port Gamble were known as fine dressers and skilled wielders of the broadax.

The "bucker" applies a crosscut saw lengthwise to felled trees before they could be "yarded." Notice the steel looping cable placed by a "chokesetter." Buckers needed fortitude, patience, and alertness for lonely work beset by rolling logs, falling timber, flying tree limbs, and flying cables.

Boss loggers wore costly black Stetson hats to signal their hard-knuckled authority. Andrew J. Pope reputedly treated visiting boss loggers to lavish entertainment on San Francisco's notorious Barbary Coast.

The Puget Mill Co., like competitors, advanced supplies and large sums to the small logging contractors, and was owed a huge sum—$ 535,000—by 1890. It was respected for paying promptly and in drafts all banks honored, and for recruiting loggers and teamsters for contractors. At the same time, company logscalers' implemented a "take-it-or-leave-it" approach to board-feet log measurements and consequent payments. Many contractors barely survived year to year. By the mid-1880s, relations between the mills and contractors were acrimonious.

Puget Mill Co. logging proceeded simply, if crudely, in the first three decades. Before cutting orders arrived, usually from the San Francisco office, loggers located the best-tapered and straightest Douglas firs with suitable grade for spars, masts, or timber "sticks." (Until loggers and mills had ways to handle them, forest giants stood untouched.) A few years after cutting a chosen tree, contractors returned to a tract to log a second or third time.

Bringing down a big fir before the 1870s required two strong people swinging single-bitted axes for up to an hour. Large trees had always been felled with axes because saws clogged on rosin and sawdust. "Swampers" sometimes first cut away thick understories or cleared the way for skid roads. A few feet up the tree trunk, fallers—called cutters, sawyers, or choppers elsewhere in the West—working in pairs cut a place for a springboard ("chopping board") to stand on. Then they undercut each side, trying to direct the fall. Danger lurked. The tree might slowly tear away or split as it thundered to the ground, killing or hurting crew members. Otherwise, loggers ignored its wake of natural destruction. The huge stump remained.

In the 1870s, western choppers switched to double-edged axes. One edge trimmed hard knots and did other dulling work. The other edge was saved for undercutting. The device shortened hours spent sharpening axes to razor edges. Axes alone felled trees until the 1880s, when crosscut saws supplemented them. After one undercut, fallers finished with

iron wedges and a crosscut saw. Sawing toppled a tree faster and more neatly and improved the direction control, as did using a later T-square device. The time it took to fall a big tree shortened by four-fifths.

One or a pair of buckers on a crosscut, armed with great patience and fortitude, laboriously cut and sawed each great log lengthwise and dodged rolling logs and falling timber. All above the first branches remained on the ground for careful burning—slash fires could spread disastrously—along with bark, broken logs,

A high climber has limbed a Douglas fir and just topped its crown. The shocked tree will swing wildly. In the era of high-lead logging, a high climber was essential in preparing a spar tree and rigging it with cables to hoist logs from the forest.

European-, Canadian-, and American-born loggers, mostly single and transient, crowded dark, smelly, vermin-infested bunkhouses throughout Northwest timber country. Cast-iron stoves heated the rooms and dried clothes.

and other debris. Cyrus Walker fumed in 1885 about a tree slaughter that wasted wood worth milling. While he approved selecting the best trees, he wanted more than the best portions harvested. Yet with Northwestern trees still so cheap, nobody had much financial incentive to impose less destructive ways.

Removing downed timber was logging's most difficult problem. Stout levers turned logs before introduction of the peavey, a lever combined with a moveable, curved hook. Yarding was by skid road and ox or horse teams to waterways. Western loggers invented the skid road. Mud or gouged ruts barred dragging over plain ground. The Northwest lacked the hard frozen surfaces depended on in the wintry New England and Lake states and in Scandinavia. Ox teams skidded a string of bulky Puget logs up to a mile until they spilled into a river or cove. In the 1870s eight oxen and ten men could haul about 100,000 board feet a week about a mile, one observer reported. Herded between boomsticks and chained in the water, Puget Mill Company logs, in eastern style, awaited a tug's cable. Or, contrary to the WPA account cited above, they awaited river drives of logs, risky to life and limb but used on northern Puget Sound until the early twentieth century.

In the 1880s steam power began to fundamentally alter western logging. The company installed saws to cut even the biggest logs. The high-capacity steam mills needed a widening circle of raw materials to keep running. Mechanization in the woods enabled loggers to cut their way away from sites near Puget Sound toward mountain crests. Narrow-gauge log trains wound deep into uncut terrains. Their Shay-geared locomotives had the power and traction to maneuver the long, heavily loaded flatcars or the wheel-mounted logs around snaking curves and on steep, winding grades. Animal-power skidding sharply declined, and disappeared altogether just after the turn of the century. Loggers no longer were restricted to less than a mile of skid road yarding.

Logging sites in the new era acquired a rough industrial appearance. Heavy equipment was needed to clear and pack down big areas for donkey engines and other equipment. Railroads clove the sites. Puffing engines shrouded nearby woods in smoke and steam. Donkeys spun out heavy cables to the hoists, derricks, and drag lines that yarded logs to streams, tidewater, or flatcars. Once the only sounds had come from the thunk of an ax, the ripping of a crosscut, the falling of a tree, and the crack of a whip as a "bull puncher" cursed oxen forward. Now the sites smelled sharply of heavy oils, rosin, sawdust, and burning waste. Loud sounds issued from the heavy machinery.

Tragically, mechanization accelerated gruesome accidents among men laboring from daylight to dusk. Lethal steel

Yarding giant Puget Sound logs was a monumental task in the nineteenth century. Here, a team of oxen hauls a log along a skid road. Short logs or poles were placed at right angles to the direction of the log movement and, in early days, were greased with dogfish liver oil. Some terrains and logs permitted horse teams.

wires might shriek through the air or hide like serpents among logs and underbrush. A log swinging from a spar tree could knock down two-foot-thick trees and pulp loggers without losing momentum. Few warnings were given or expected. Yet the workers labored on: Dissatisfied loggers or those unable to keep up knew that they were easily replaced.

By the 1910s operators adopted the rigged spar tree and donkey-powered high-lead, with their intricate cableways. The high-lead yanked logs into the air to the yarding area. Awed crews watched the preparation of spar trees. A "high climber" went up 100 to 200 feet, lopped off 50 to 60 feet of top, and rode the shocked, wildly whipping spar eight or more floors above the ground before installing a huge block and pulleys. Schools were dismissed to watch the acrobatics.

Loggers still felled the best timber and left lower-priced grades and unwanted species strewn on the ground. Crews burned the spar and the 3 to 5 feet of litter and debris. A tree stub stood here or there. Nobody expected to log high-lead tracts again.

Western forestry output skyrocketed with mechanization. Donkey engines reportedly more than halved log removal costs. Hauling logs by rail ten to fifteen miles to landings cost about the same as yarding them a mile by oxen. Taking logs cheaply and efficiently, however, required increasingly large capital investments for donkey engines, high-leads, and—not mentioned in this sidebar—short-haul logging railroads.

Railroads in the United States left mills to build their own narrow-gauge lines. Washington's first logging line appeared south of Olympia in 1881. The Puget Mill Company began extensive railroad building in the Olympic foothills in 1885. Flatcars on its first narrow-gauge in Mason County tipped logs into Hood Canal and elsewhere into rivers.

The capital needed for logging locomotives and cars, leagues of track, countless bridges and culverts, and more facilities and employees ultimately forced sawmills to handle logging directly. Entering its own tracts, the firm set up logging camps in the 1910s. Top executive Edwin G. Ames predicted in 1912 that the immense investments required for logging railroads, Lidgerwood outfits (steel substitutes for spar trees), and donkey engines doomed independent loggers. For a while these developments did drive them from the woods. Gasoline engines brought them back.

CHAPTER THREE

GROWTH

1881-1924

AT THE TOP

IN 1881 A CONFIDENT WILLIAM H. TALBOT TOOK HIS APPOINTED PLACE AT THE HEAD OF A major firm in a key western industry. Family properties, wealth, and connections gave him a yet untested authority. Although only twenty-one years of age, he had something of the late captain's commanding presence. The younger Talbot was reserved, deep-voiced, and, at almost six feet in height, well built. He was unbending in manner and carriage and always impeccably dressed.

Business dominated Talbot's life. Like his predecessors, he proved to be a hands-on manager devoted to details. Everything from daily expenditures to fire-protection techniques received minute attention. Neither difficult journeys to the mills nor long hours in the office unsettled or bored him. Above all, Talbot wanted family members to remain in charge, and a son to eventually serve as successor.

Something of the shrewd Yankee trader clung to Talbot. When foreign customers demanded lumber free of sap, he ordered that care be taken with such cargoes until vigilance was relaxed in overseas ports. At that time, he instructed, the proportion of higher-sap lumber shipped should be gradually increased.

Though tough and persistent, Talbot could be defeated. Some competitors learned how to best him in the rich Hawaiian market, but only after decades of trying. When a Seattle rival threatened to undercut prices, Talbot vowed a fight to the finish. He was not, however, fond of competition. His solution to the overproduction that continued to plague the industry was cooperation with competitors. He took on major roles in a new series of price-fixing pools and trade associations.

On Puget Sound not much escaped Cyrus Walker. Prowling the mill dressed as a workman, he epitomized the thrifty, teetotaling, hard-working Yankee tinkerer. Walker and his brother William, a veteran hand in his own right, enjoyed nothing more than modifying equipment. Nothing was ordered done by others that Cyrus Walker could not do himself. Waste, whether of material or time, pained him greatly. In slack periods Walker toiled at various menial chores, chipping mortar from old bricks or repairing chimneys. To save money, he picked up stray nails. At night he worked on the accounts and correspondence with the San Francisco office. If he did not go to bed physically tired, Walker said, he had failed in his duty. While Talbots and Popes resided in San Francisco, this austere man rarely left Puget Sound, at least before 1906, when an industrial accident and the onset of senility compelled his retirement.

> *"Partners who have agreed to work together because of shared goals have a very different relationship than do those of their heirs, who find themselves working with people they didn't choose. I think that helps explain why the transition from one generation to the next in a family business is very difficult, and why many enterprises don't survive this period. Pope & Talbot was unique in that Andrew J. Pope died when his son was only fourteen, whereas Captain William C. Talbot lived three more years and put a fully grown, twenty-three-year-old son, William H. Talbot, in charge. William H. Talbot was a very successful businessman who didn't need the help of a younger cousin, my grandfather George A. Pope. He was able to rule the family business with a strong hand. Ironically, though, this advantage also carried with it the seeds of future trouble: Because he held so little company power, George Pope developed interests outside of Pope & Talbot. He became an active investor in stocks, bonds, and real estate. He was a prominent figure in San Francisco society. Later, this lack of involvement in company affairs almost led to the end of Pope & Talbot."*
>
> —Peter T. Pope

In 1887 the Walkers quit Port Gamble for a large wooden mansion, Admiralty Hall, in Port Ludlow. Walker never liked palatial living, and he proceeded to turn the grand house into an extension of his work. The baronial showplace, which contained a great room for the entertainment of visiting captains and customers, became instead a place of business. The entrance hallway, leading to a grand staircase, resembled a ship's interior. Carved black walnut furniture from New England and Europe and trinkets from around the world filled the rooms. A brass cannon saluted arriving vessels.

The lobby of the Admiralty Hotel, seen from the billiard room. Before becoming the hotel, Admiralty Hall was Cyrus Walker's palatial residence where he and his wife entertained customers, captains, and the political and business figures who composed the "sawdust aristocracy."
(John D. Cress, photographer)

36 GROWTH

The former Admiralty Hall, Cyrus Walker's mammoth home built in 1887, commanded a site overlooking the mill and the harbor in Port Ludlow. It became a hotel, then was torn down. *(Webster & Stevens, photographers)*

Edwin G. Ames took Walker's place at Port Gamble. Six years earlier the Talbots had sent this twenty-five-year-old cousin west to reduce Walker's burdens. Ames was born and raised in East Machias, Maine, where his father was a Pope lumber manager. What he did not already know about sawmills he quickly learned on the Sound. A fervent Republican and advocate of low taxes, Ames hated troublemakers in general and unions in particular. Like Talbot, he believed in making, and actually honoring, private agreements among fellow mill owners.

Ames was another six-footer, smart, ambitious, and assertive. Reserved in manner, he seemed overly formal and even pretentiously aristocratic to rough-hewn observers. Nobody ever caught him scavenging nails on the mill floor. Mastering each of the operations, he advocated more efficient methods. As much of each tree as possible had to be utilized, Ames urged, rather than burned as waste. He wanted to diversify, and stressed highly manufactured products over low-profit green lumber. Do not hire relatives, he advised, unless they are willing to start at the bottom and work their way up. Under the Ames regime prospective salespeople first spent a year working in the mill.

Ames greatly admired Walker, who liked and trusted the younger man in return. They shared a strict work ethic, New England customs, and political views. Neither made decisions in a hurry. In 1888 Ames married Walker's niece Maude, the daughter of William Walker. Her parents had entertained their future son-in-law for seven years at Sunday dinners and holiday celebrations. Ames joined Cyrus Walker in real estate speculations near Seattle. As that city prospered, he became a land-based millionaire.

William H. Talbot's first seven years at the helm coincided with one of the most prosperous periods yet experienced on the Pacific Slope. Operators later looked back on the 1880s as "the good old times." While the population of the United States rose by 52 percent during the decade, lumber production mounted by 94 percent.

The "big sticks" are dumped from the company's narrow-gauged railcars into Puget Sound in 1918. Almost all nineteenth- and twentieth-century logging entailed using waterways closest to the fallen timber to get logs to the mills.

The Talbots, Walker, and longtime associate A. W. Jackson opened new California lumberyards in partnership with the Hooper brothers, the owners of redwood mills, between 1885 and 1890. The yards provided the Puget Mill Company and the Hooper operations with outlets in the San Joaquin Valley and in southern California. Three Talbot-Hooper yards were also opened in San Francisco, one inside Pope & Talbot's own huge facility at Third and Berry Streets. "Oregon pine [Douglas fir], Redwood, Shakes, Laths, Shingles, Posts . . . and Building Material of Every Description" were marketed. Some of the yards also dried, planed, and remanufactured timber from the mills.

The youthful Talbot had good reasons for confidence. His three mills were great producers and effective traders. The company had plenty of money. Land, machinery, and ships and tugboats were readily available for acquisition. The cost of sailing vessels dropped, as steamships gained in general favor. Bankruptcies made mills and equipment quite obtainable. And for one last time, one of the biggest private timber owners on the coast was able to secure cheap government land.

Talbot's top executives were highly capable. Family relations fostered a sense of unity and loyalty within the company. Tight-knit, trusting, and often affectionate relations prevailed among the families. Board member Charles F. A. Talbot handled customers, captains, and endless details. Brother-in-law Cyrus Walker and cousin Edwin Ames ran things on the Sound. And the president's younger brother, Frederick C. Talbot, rapidly learned the executive ropes.

Morale was also high among the skilled workers. The men at the mill prided themselves on handling

> "In San Francisco, Pope & Talbot is still thought of as a steamship company. For more than a century company shipping operations were based in the Bay city. Until Oakridge opened in 1948, all of our mills were cargo mills, not rail mills. Owners have always given ships a lot of attention. Many of the most influential Pope & Talbot executives had shipping backgrounds. In the twentieth century the steamships typically did well for the company when the economy was strong, as opposed to housing, which did better in a weak or recovering economy, when interest rates were low. Steamships kept Pope & Talbot a diversified company, and not just a lumber producer."
> —Peter T. Pope

the best and biggest logs in the West. E. C. Fitzhenry, a company sawyer for three decades, bragged about cutting 16-inch continuous vertical grains in logs perfectly clear of knots or other defects. The ruler-straight timber became 100- to 120-foot masts.

The mills and logging camps had, for the first time, a more than sufficient supply of workers. An army of casual laborers arrived on Puget Sound by the new transcontinental railroad during the 1880s. Among the many migrants from northern Europe were experienced Scandinavian loggers and mill workers. Other newcomers were veterans of the cutover forests of the Great Lakes states. Many workers, to the consternation of employers, were restive and interested in unionization. Walker complained that the contemporary man lacked determination and grit. In his view slackers, whiners, and other unmanly sorts insisted on fine accommodations and food and top wages.

VILLAGE LIFE

NEW MILLS AND A DIFFERENT SORT OF WORKFORCE ALTERED THE NATURE OF THE COMpany's towns. Increasing numbers of foreigners took up residence. From 1880 on, the percentage of New Englanders declined. Walker expected single newcomers to furnish their own blankets and dine at the cookhouse. The immigrants were welcome in the clubs, churches, and saloons, at Saturday-night dances and summer baseball games. Unlike in Ludlow or Utsalady, Port Gamble's population did not grow. It had just over four hundred residents in both 1880 and 1890. Many held fast to the New England style. Old-timers boasted of the well-dressed citizenry, the good water, the neat thoroughfares, the well-stocked library, and the Masonic and Odd Fellows chapters. The entire community assembled for holiday celebrations and clean-up days. Regular steamer service to Seattle lessened the onetime sense of isolation.

Class and ethnic divisions nonetheless sharpened. Company hotels and mess halls maintained a strict hierarchy. At Port Ludlow the managers and office force ate four to a table in the Tyee Room, choosing among several menu items. Tallymen and foremen sat at tables in a separate dining room. All other employees consumed whatever was served in the big mess hall, presenting a metal check at the door and occupying assigned seating. Ten to twenty workers quit every day, but an equal number signed up to take the

THE COMPANY BARK ATALANTA GOES DOWN AT SEA

"WE LEFT THE MILL AT PORT GAMBLE, IN TOW OF the tug Tyee, *December eighth [1890]. All went well until I put sail on off [Cape] Flattery; the vessel then began making water, but, being lumber-laden, I did not think it worth while [sic] to go back. We got down as far as the mouth of the Columbia, and the wind shifted from northwest to southwest, with snow squalls, and it was then that the forty years that the* Atalanta *had been afloat told on her with fearful effect. The sails all blew away on the night of the thirteenth, and soon after the heavy deck load of 80-foot timbers broke adrift, and on the morning of the fourteenth the fore and main mast went by the board, the foremast smashing the long-boat, destroying our means of leaving the ship. . . .*

About noon of the fourteenth the vessel commenced to break up, and about 3:00 P.M. she parted just abaft the main hatch, leaving fourteen of us on the after-house, with nothing to eat or drink . . . Night began to set in, and a night in the month of December off Vancouver Island [British Columbia] is a long one, even when one is comfortably situated. The mizzenmast went shortly after daylight and took nearly one-half of our limited raft. Through all that day and the next night the sea was making a clean breach over us, but on the morning of the sixteenth we sighted land, which was a relief even though it was far away. The steward, John W. Wilburn, became temporarily insane at noon on the sixteenth, the first officer's leg was broken, and all hands were inclined to feel despondent. We had full made up our minds that we would either be dead or ashore before morning, as we were all very badly chilled. . . . [So] we passed the longest night I have ever experienced. . . .

[By morning] the rudder had become jammed with a lot of the deck load, forming quite a raft, and [it and the after-house, with men clinging to both, were swept by an eddy] directly on shore . . . having drifted 170 miles on the raft in four days and four nights . . . without losing a life. The Indians were very kind to us, and we were taken to Victoria by the sealing schooner Katherine."

—Captain Frederick Thrasher, from
Lewis and Dryden's Marine History of the Pacific Northwest,
ed. E. W. Wright (Portland: Lewis & Dryden Print Co., 1895)

Lumber-carriers were nautical workhorses. Outbound from Puget Sound, lumber, spars, railroad ties, ship decking, beams, and mining timbers jammed the holds and towered on the decks. On the second—or return—legs company sailing ships, unless in ballast, normally carried commercial bulk cargoes.

Build large for Puget Sound, manger Josiah P. Keller had advised his Maine shipyard partner, Charles Foster, in 1861. Vessels ought to transport 250,000 to 400,000 board feet. Larger mill production after the Civil War demanded larger vessels. Million-or-more-board-foot cargoes to Asia, Australia, or South America were common. Few West Coast mills could load this amount—a company advantage because captains preferred one cargo to calling in at several mills.

The company's fleet usually sailed without incident, but never without risks. Lost seafarers and vessels marked its New England heritage. News of lost, crippled, and burned seafarers arrived all too often. Captain William C. Talbot's brother John, age twenty-one, disappeared at sea. Steamers, which the firm refused to buy, carried their own perils. Within sight of the Port Gamble mill, one steamer horrifyingly exploded in a rush of smoke and flames, killing all aboard.

To company executives, voyages inevitably meant risks. They tried to lessen them with good plans and sound vessels, skilled sailing and longshore crews, fair-minded captains, government-financed harbor and coast improvements, and insurance. Andrew J. Pope in February 1861 was one of twenty incorporators of the California Mutual Marine Insurance Company, the state's first insurer.

William H. Talbot and his brother Frederick were known to care about both crew and ship welfare. They necessarily left most decisions to captains. The brothers rode in fine carriages along San Francisco's Embarcadero in search of likely looking sailors for employment, then delivered them to company ships. Their captains usually supplied themselves from the Barbary Coast, Puget Sound's Little Barbary Coast, and similar seafarer hellholes.

Crew and command incidents troubled maritime life. Angry crews occasionally set cargoes afire in port, or refused to sail. Off South America, as a company long-voyager lay becalmed for days, her captain refused to lose more time by going off-course for food and water. The crew grimly fished for sustenance and caught rainwater from passing squalls until the winds revived. Incompetent, tyrannical, and drunken officers and crewmembers as well as exploitative boardinghouses, brothels, and labor providers cursed maritiming. Even the best-run ships offered seamen poor conditions and pay. Only the coal industry had more accidents than the American maritime industry in 1900.

Nature and geography imposed tolls. Storms, groundings, rocks, and accidents damaged or sank windjammers. Before the Civil War each round-trip to New York or Philadelphia required a year and meant rounding perilous Cape Horn twice. Its icy winter reaches could tear at bodies and doom sailors and vessels. Elsewhere, by 1856 company ships had capsized twice under William C. Talbot, once as survivors vainly threw off deck lumber. In January 1881 gales battered the *Rainier* into a drifting derelict and killed her master. Lashed to the poop deck, her half-starved crew survived a twenty-one-day ordeal. Two sister ships simply disappeared at sea. While under tow, a bar took the nineteen-year-old *Bonanza* in 1894.

Cargo size and storage held keen interest on lumber vessels before Plimsoll lines (loadlines) were required on hulls. At dockmaster orders in calm summer months protesting masters would be towed into Puget Sound with their decks almost awash, in the fashion of Maine lumber ships. Although it was difficult to sink one holed in the bottom, a waterlogged lumber vessel was almost unmanageable.

Captain Thrasher's 824-ton *Atalanta* and 600,000 board feet of lumber and laths were lost. Eleven months later a gale crippled the company's *James Cheston*. The crew threw off most of her deckload. The pumps failed, and the bark, down by about 14 feet, was towed back to Port Gamble. After fixing leaks, "Captain Plumb thinks it would not be safe to put any deck load on her," Cyrus Walker told Pope & Talbot's office on November 16, 1891. "And besides if we did," he continued, "it is very doubtful if we could get any of the crew not in the vessel to go, therefore I thought it best to send her to sea without a deck load." And so it sailed.

Plumb's 995-ton bark, launched in 1855, was almost as old (1851) as Thrasher's lost command. Once repaired, she arrived without incident. Frederick C. Talbot suggested to Walker on October 5, 1894, that their many elderly sailing vessels be sold before all were lost. (Both Walker's letter and Talbot's letter are housed in the Ames Collection, University of Washington.) Instead, the frugal owners maintained windjammers despite their obsolescence beside the steam-powered, steel-bottomed vessels now sweeping the oceans. It wasn't until the end of the century that the *James Cheston* was scrapped.

(left) Puget Mill Company, Port Gamble, Washington, 1898.

(right) Two rows of simple worker housing border the busy Port Ludlow mill and harbor, 1912. Farther from the mill lies Executive Row and other better company housing.

places of the departed. Transient loggers and sailors made up a substantial minority of the total population, especially when the wharf was crowded with vessels.

Most of Port Gamble's single men, except for the Chinese, were housed close to the mill in cabins, bunkhouses, or hotels. The remainder rented cabins on the outskirts of the village or nearby farms. Home-owning Maine natives often rented rooms to fellow Pine Staters. At both Gamble and Ludlow the company, always striving to develop the most stable workforce possible, loaned money to married men for the construction of houses.

Following the early-twentieth-century expansion of the mills, unmarried men often roomed at the Puget Hotel Annex in Port Gamble. Chinese laborers, cooks, and laundry personnel always lived apart, either communally or alone, from other ethnic groups. S'Klallam workers continued to reside across the bay at Little Boston. The firm built modern homes in the older residential quarter for management, the office force, and skilled mill employees. The big house occupied by Ames, the Puget Hotel, which was opened to summertime tourists, and a community hall graced this section of town.

Second-growth forest and the cemetery divided upper crust and worker Port Gamble. The latter district made do with small cottages, unpaved streets, and wooden sidewalks. Murphy Row anchored one end of the town and "New England" the other. When Port Ludlow closed down in the 1930s, the disassembled houses that were put back together at Gamble became known as Little Ludlow. By then the older laboring section had a church, a post office, and even cement walkways. As in the past, however, children from the opposite side of the tracks did not associate with one another.

CHANGES

THE MODERN UNITED STATES OF BIG INDUSTRY AND BIG CITIES EMERGED IN THE FIRST years of the twentieth century. Railroads and, to a lesser extent, unions became important factors in company history. Far West lumbering was no longer isolated from national economic currents. The northern transcontinental railroads finished during the 1880s and the 1890s completed the integration of the Pacific Northwest into the rest of the United States. Southern and western sawmills now competed with one another in the

Port Gamble's meticulously maintained machine shop, 1918. Machine shops were vital to keeping sawmills running. From them emerged inventions or improvements to many of the nation's sawmills.
(Webster & Stevens, photographers)

Lumber Manufacturing plant, Puget Mill Co., Port Gamble, Washington, 1918.
(Webster & Stevens, photographers)

eastern and midwestern markets. National lumber prices boomed. Western mills were fully exposed to the country's long-term deflationary trend. Low prices drove mills into overproduction, price-cutting, and cartel organization. Bankruptcies increased.

In Washington the coming of the Northern Pacific Railroad in 1883 generated a growth in population and, as Cyrus Walker regularly lamented, a sawmill construction craze. Tacoma became a rail mill center. Manufacturers opened plants on Grays Harbor and Bellingham Bay. Investors interested in the making of lumber and shingles founded Everett. Railroads connected these newly important places to domestic markets closed to the cargo mills. No railroad built into Gamble, Ludlow, or Utsalady. Mills "handy to the Rail Road all have the best end of the biz," Ames pointed out, "as they will be able to market lumber & waste that we are obliged to burn."

Railroad lines—the great transcontinentals, their assorted branches, and the numerous logging roads—covered the Northwest. Rail opened previously inaccessible forests to logging. Great Lakes operators, having depleted their pineries, cruised the region for timber and mill sites. As early as 1888, Talbot seriously considered the idea of finding a buyer and selling the business. That year the Puget Mill Company failed to return a profit for the first time in its history. Cyrus Walker wanted to unload "this elephant." At more optimistic moments, however, Talbot thought that the new rail mills would avoid exploiting his traditional water markets. Rail shipments, he reasoned, would be so lucrative that the newer mills would leave California and the Pacific Rim alone. They did not.

In another major change of the 1880s, labor organizations imposed themselves on Northwest lumber interests. "To Walker, Ames and other worried lumbermen," writes historian Robert E. Ficken, "the future carried visions of crazed and besotted unionists wresting away the property of virtuous Americans." By 1886 a longtime campaign by the Knights of Labor for the ten-hour day and Chinese exclusion had driven Chinese residents and workers from several Puget Sound towns and many of the mills. The Knights quickly lost force, but wage demands by the sailors union continued to vex Walker. "I don't know where this increase in wages is going to stop," he complained. "Everybody seems to want more money for less work."

Attitudes toward labor hardened. "There is a strong inclination among all the employees," Walker observed, "to become high toned as to how they shall live, the time they shall work, the amount of work they do & the pay they receive, which makes things run harder than in times past." Talbot agreed. The threat or actual use of nonunion replacements often succeeded. The sailors union, Walker reported in 1892, "fear us more than any other house on the coast, & we can get more concessions from them than the rest." Whenever the firm attained a truly commanding position in the market for labor, Walker recommended that old scores be settled. No union, Ames declared in 1904, had any right to interfere with management control over hiring and firing, the hours of work, or the rate of pay.

Talbot began his half century at the corporate helm as a cautious innovator. Change followed mainly along the lines laid out by his predecessors. He reinvested profits despite occasional family pleas for higher payouts. The Puget Mill Company distributed only ten dividend payments between 1874 and 1914. Borrowing was avoided. Surprised in 1888 by $20,000 in overdrafts, Talbot hurriedly deposited money because "we don't like to pay interest—the Co. never having done it." Despite the inherent problems, he continued his father's preference for pooling arrangements. With Talbot in the lead, thirteen Washington and Oregon mills formed the Pacific Pine Lumber Company in 1886.

Talbot was confident that the firm could hold and expand markets against other cargo mills and the multiplying rail plants. Like his forebears, he concentrated on perfecting sales and distribution networks. In the premier California market he opened new Pope & Talbot and Talbot-Hooper yards. The mills aggressively pursued export business, profiting from the emergence of plantation agriculture in Hawai'i, the spread of mining and grazing in Australia, and railroad construction in Chile, China, and Peru.

For as long as possible, Talbot and Walker resisted making costly changes. Many of the other western mills, they noted, had gone bankrupt instituting expensive and untimely improvements. The firm continued to rely on buying and upgrading equipment from failed mills and on individual mechanical ingenuity. Company mills retained rotary saws despite the availability of superior band saws. Kilns were not installed until after 1890, despite customer demand for better material than wet green lumber.

Pope & Talbot also continued to rely on sailing vessels, maintaining the biggest fleet on Puget Sound, a dozen and more barks, barkentines, and ships in the 1870s. Early in the 1880s Talbot ordered larger and more efficient sailing craft, designed specifically for the handling of wood products. He also bought older windjammers for renovation; company mills supplied the labor and materials. In short, Talbot stubbornly refused to transfer the company flag to the more efficient but costly iron- and steel-bottomed steamships then commonly used in world trade. With improved boilers and engines, an iron vessel carried 150 percent more cargo weight than a wooden competitor. By 1900 the firm owned, or controlled as part-owner, thirty-one sailing barks, barkentines, schooners, and ships. When sending a 3-million-board-feet cargo to Asia, it had to charter a foreign steamer.

Talbot began to mechanize logging, at least indirectly. To avoid the high cost of company logging camps and modern equipment, he loaned large sums to logging contractors. Donkey engines, high-lead yarding, and, beginning in the Olympic foothills in 1885, logging railroads fundamentally transformed the industry. Output expanded as loggers operated over more extensive and difficult sections of forest. No longer dependent on skid roads, cutters moved inland and to higher elevations. The company thereby helped create, primarily through financing, a new era of logging.

A heavily rigged lumber vessel loads over the stern at Port Ludlow.

THE GRAYS HARBOR COMMERCIAL COMPANY

The lumber mill and burner at Grays Harbor Commercial Company in Cosmopolis, 1912, when it was controlled by Pope & Talbot executives and associates. Starting in the 1890s, a long freight train could load beside the mill. (L. A. Dix, photographer)

SETTLERS AND LUMBER PEOPLE LARGELY IGNORED southwestern Washington before well-financed Great Lakes lumber interests began exploiting the Grays Harbor area's vast timber resources in 1882. The new owners speedily made Tacoma and the cluster of Aberdeen, Cosmopolis, and Hoquiam major rivals of cargo mills at Ports Blakely, Gamble, Ludlow, and Madison. Free-spending harbor developers introduced up-to-date sawmills, logging, and ships and cut timber without regard to the future. They Tried to Cut It All, local historian Edwin Van Syckle entitled his history. Customers got rough lumber for about what the Puget Mill Company paid for logs.

Located on the west coast of the Olympic Peninsula, Grays Harbor significantly reduced the Washington lumber industry's ties to San Francisco. Owners, living on the harbor, were not dependent on San Francisco capital and depended less on the California market than did Puget Sound mills. Railroads reaching them in the mid-1890s largely turned harbor plants into rail mills, thus allowing them to heavily penetrate Midwestern markets.

William H. Talbot soon felt forced to confront their challenge. He had the Pacific Pine Lumber Company, in which he played a leading role, make a defensive investment. This cartel bought and expanded a Cosmopolis sawmill in 1888. It constructed a company town, rough camps, logging railroads, prodigious barns, many production units, and great wharves. Operations integrated from timber to tidewater and new rail links made the Grays Harbor Commercial Company, as it became known, one of the state's biggest manufacturers by 1892.

The Puget Mill Company and some of its executives came to own a stock majority. A. W. Jackson, former Puget

Mill sales executive and an investment associate of Talbot and Walker, headed it for a time. William's son, Frederick C. Talbot, learned management on the harbor. He was renowned for signing Alaskan fish packers to a long-term contract for shooks, the sawn or split wood pieces used in box construction.

A continually profitable Grays Harbor Commercial Company dwarfed the Puget Mill Company. It was the area's largest manufacturer. In the new century, twelve hundred employees cut some 600,000 board feet a day. It shipped more than 39 million feet annually, 33 million of it by rail. Trains loaded from a half-mile-long shed. In San Francisco the firm maintained a box factory. Pope & Talbot or a Talbot-related lumberyard provided San Francisco sales facilities; after 1912, Grays Harbor shared Pope & Talbot's yard in Seattle.

In and around what was called "dirty old Cosmopolis," the firm constructed a wooden farm tank factory, a sash and door factory, dry kilns, one of the industry's biggest planing mills, and two large shingle mills. Managers created a brisk market in specialty products like staves for silos. They ran a hog farm, slaughterhouse, bakery, company-scrip store, several bunkhouses, and logging camps. Employees given twenty-minute meal breaks almost trampled one another entering the six-hundred-seat mess hall. Cornmeal mush was the staple. Mess hall rat-catching contests enlivened Sundays. Meanwhile, the mansion of third manager and part-owner Neil Cooney was the scene of lavish, all-night parties. Pope & Talbot and Puget Mill Company executives lived there during periodic visits.

The firm had a storied or notorious character throughout the West. Nicknames abounded: the Old Folks Home, Neil Cooney's Empire, the Western Penitentiary. It was said that if you went to hell, you would find somebody who had worked at Cosmopolis. Port Gamble filer Will Thompson called Cooney a "hard hearted son-of-a-gun [who] ground his men down to the last nickel." Ralph W. Andrew's This Was Sawmilling likened the mill to a "feudal estate where a man could wrestle lumber as long or as little as he liked at the lowest possible wage," knowing he had bug-infested accommodations and bosses skilled in cutting logs as cheaply as possible. Key executives and those with top skills received good pay.

Individuals from all walks of life filled Grays Harbor's ranks. Unlike the earlier Puget Mill Company, the firm maintained a come-one, come-all policy. Butte, Minneapolis, Portland, San Francisco, and Seattle labor suppliers sent a certain number of workers every week. Age, race, nationality, language, and physical or mental condition were irrelevant. Hard times and personal problems brought people already resigned to receiving little more than food and lodging for a month or two. Foremen ignored the constant turnover so long as travel advances, recruiters' fees, and dollar-a-month medical fees were worked off. Cosmopolis had no labor shortage. The Far West, especially before the First World War, was criss-crossed by an army of itinerant laborers and by machine-tending operatives wanting to improve their skills and status.

A boy working at Grays Harbor in 1903 earned board and a dollar a day. Egbert S. Oliver, in the Pacific Northwest Quarterly in January 1978, recalled his 1917 sojourn. For a sixty-hour week the fourteen-year-old received $40 a month—around 16 cents an hour—and ample but inedible mess hall fare. He rented dismal lodgings. His work consisted of precisely stacking the wooden box ends that were spit at him from a machine at a rate of sixty ends a minute. Young and nimble, Oliver learned to respond to the machine's timing. If he lost count, the ends slammed into his thigh. "I was bruised enough to be black and blue most of my working days" but "learned to live and breath [sic] with the machine, adjusting my conscious and subconscious time mechanism to that count of five."

The operation kept producing even when the Industrial Workers of the World closed nearly all Northwest mills and timber camps during strikes in 1912 and 1917. Frederick C. Talbot assured Cooney in April 1912 that Cooney's strong-arm tactics were the "right thing." Armed guards, electrified fences, and fierce dogs surrounded Oliver's plant.

Conditions changed radically during the next decade. The mill cleared all of its own timber, grew dependent on log buying, and wore out its facilities. Cooney's broken-down empire could not be sold during the lumber recession of the 1920s. As the Great Depression began, the company dissolved in 1929.

Burner, Puget Mill Co., Port Ludlow, Washington, 1918. To get rid of waste and prevent mill fires, burners ran around the clock in Port Gamble and Port Ludlow until the 1920s. Smoke (seen for miles), ash, and cinders wafted into the towns. Later generations found uses for the waste. *(Webster & Stevens, photographers)*

REORGANIZATION

EVER CAUTIOUS, WILLIAM AND FREDERICK TALBOT WERE PIONEERS IN AT LEAST ONE area: The young executives terminated the firm's traditional seat-of-the-pants management style. In 1888 and 1890 the brothers untangled cross-ownerships, reorganized units to streamline management, and clarified responsibilities within and between the several concerns. A welter of partnerships that had been formed over three decades were transformed into corporations linked to Pope & Talbot and the Puget Mill Company. Reorganization enabled the families to spread their timber holdings among new legal entities, sidestepping recent Washington State legislation restricting timber ownership by any single firm.

Pope & Talbot was the first western lumber company to adopt systematic business management methods. Its executives henceforth required meticulous accounts and methodical reporting. Correspondence had to be typed, not handwritten. Bookkeepers were hired to conduct unannounced field inspections. The Talbots wanted to know exactly the board foot potential of trees, the volume of logs in the pond, and the exact amount of money owed by logging firms and other debtors. How many pennies an hour per thousand board feet would a new machine save? How many pennies a pound were gained by selling sugar to the hotels and mess halls from the company stores?

General manager Walker headed six Puget Mill Company divisions and four geographic subdivisions. In 1890 the new regime discovered that loggers owed $535,000, a revelation triggering additional reorganization. The company incorporated five separate affiliates under California law. The Puget Lumber Company primarily handled the mills and some sales. The Puget Shipping Company directed export sales. The Puget Trading Company ran company stores. The Puget Commercial Company owned the controlling interest in the Grays Harbor Commercial Company, a firm established by the Pacific Pine Lumber Company under prodding from Talbot. The Puget Mercantile Company existed only on paper.

In 1891 Talbot established the Puget Sound Towage Company to tow logs. He also joined three nearby Sound mills in forming a noncompeting pool, the Puget Sound Tug Boat Company. The Rainier Investment Company was incorporated in California in 1893, under George A. Pope, to hold and manage Washington real estate owned by family members. In return for various fees and commissions, Pope & Talbot took charge of Puget Mill Company sales and finances, the affiliates and subsidiaries, and the Rainier Investment Company. In 1901 William and Frederick Talbot turned the Pope & Talbot partnership into a California corporation.

HARD TIMES AND BEYOND

THE PANIC OF 1893 INITIATED THE WORST DEPRESSION THUS FAR IN THE NATION'S HIStory. A recession demoralized Northwest lumbermen before the panic. Worsening conditions eventually ruined the Pacific Pine Lumber Company. Members dropped out and 11 million board feet choked the pool yards. At the end of 1890 the Puget Mill Company placed Port Ludlow on standby status and indefinitely shut the Utsalady mill, in return for subsidies from still operating participants in the combination.

The depression cut a deep swath through the United States. Washington State experienced massive business failures, mass unemployment, and social unrest. Population growth slowed. As prices tumbled, resource-based firms staggered from crisis to crisis. The new rail trade stalled; three-quarters of the shingle plants closed. Desperate lumber concerns offered themselves for sale, with no takers. Lumber was dumped on every conceivable market. Historically important sawmills shut down forever at Port Discovery, Port Madison, and Seabeck. Utsalady never reopened.

Sorely tested, the firm tied up ships, reduced wages, laid off workers, and stripped Utsalady of usable equipment. Outdated machinery and aged vessels continued in service. Timber poachers invaded company land. Squatters were so numerous that mill officials, though fearful of the risk of fire, charged dollar-a-month rents.

Port Gamble continued to produce for the Hawaiian market. A hearty new trade opened with South Africa in low-grade building and mining material. Talbot secured his first bank loan at this time, an alarming violation of familial doctrine. When the Pacific Pine Lumber Company proved ineffective, Talbot and twenty other mill owners formed the Central Lumber Company in 1895. Before closing down in 1897, this latest pool grew to include sixty members, all but three of the exporting mills in the United States and Canadian Pacific Northwest.

The company reopened Port Ludlow in 1897, as William McKinley's presidential inauguration somehow convinced Walker that recovery was under way. Republican mill owners credited McKinley for the return of good times, just as they had blamed Democratic predecessor Grover Cleveland for the depression. Edwin G. Ames, in particular, linked the company's welfare with the continuation in office of a business-oriented Republican party. The forest-products industry continued as a mainstay of Washington State's majority party through the election of Franklin D. Roosevelt in 1932.

The year 1897 was good for the company's mills; 1898 was better and 1899 better still. The Alaskan gold strike enriched Washington, although lumber had already glutted the Skagway market by the time Talbot had sent his first chartered steamer to that port. Wood prices increased nationally as the depression wound down. Eastern buyers once again looked for timber, mill sites, and mills. Washington rail shipments doubled between 1899 and 1901.

Overseas manufacturers and railroads demanded American lumber and related forest products. The Alaskan Packers Association placed heavy orders. Pope & Talbot boosted sales in Australia, China, Korea, South America, and in particular in Japan. Hawaiian sales surged after 1898, in response to American annexation and the war with Spain. The offshore market, happily, absorbed low-grade material, since California now demanded the better "clear," or nearly defect-free grades.

Rival lumbermen were interested in actually acquiring the company. A representative of unnamed Midwestern capitalists approached Walker in March 1899, asking for detailed inside information about the firm. Walker and Talbot agreed that the man was on assignment from Frederick Weyerhaeuser's syndicate, currently rumored to be on a buying spree. The Weyerhaeuser group had already bought 45,000 acres on northern Puget Sound. Pope & Talbot was receptive to selling. We "must make our pile," Walker wrote the San Francisco office, "and be satisfied to lead quieter lives."

Convinced that timber was bound to increase substantially in value, Talbot and Walker demanded too high a price. The Weyerhaeuser forces soon bought 900,000 acres from the Northern Pacific Railroad, immediately becoming the nation's second-largest private timber-holder. With additional purchases included, Weyerhaeuser bestrode the Northwest industry. At that time the control of timber, rather than of sawmills, was still the key to success.

Walker predicted in 1901 that in the near future the biggest timber-holders would control stumpage prices. Charles F. A. Talbot welcomed such a result. The fewer the owners, he advised, the easier the task of

The sawing floor at Port Ludlow, the second-largest Puget Mill Company mill at the time of this photo, 1918. The mill had been modernized in 1904. *(Webster & Stevens, photographers)*

dictating prices. The Weyerhaeusers and other Great Lakes investors added tens of thousands of acres, driving up timber values to new heights. The Talbots and their associates offered to sell Weyerhaeuser their controlling interest in the Grays Harbor Commercial Company, but the new industry leader refused to pay the asking price.

In 1903 the Port Blakely Mill Company, previously second only to Puget Mill among traditional manufacturers, passed into Midwestern ownership. The leadership of the Northwest industry shifted decisively from San Francisco to Saint Paul and the Great Lakes. Only the family-owned Pope & Talbot firm had survived from the pioneer era. It remained on the market. Potential buyers were drawn not by its elderly sailing ships and run-down mills, but by its standing timber.

VOLATILITY

THE COMPANY PROSPERED BETWEEN 1897 AND 1907 DESPITE INFLATION, BETTER-SITUATED competitors, and other difficulties. The speculative mania in timber and mills unleashed production. Marginal operators ran hard to pay for their high-priced timber. Larger mills had the raw material facilities to substantially increase output. Between domestic and international demand for wood and its own conservatism and careful management, Pope & Talbot earned substantial profits. Ships crowded the Gamble and Ludlow wharves, annual shipments averaging 87 million board feet.

In 1906 the San Francisco earthquake triggered one last boom for the old-time mills. Five hundred city

A Knox Martin Tractor, now in the Smithsonian Institution, will draw 10 tons of Pope & Talbot's 90-foot-long "sticks" along San Francisco's narrow and hilly streets in 1913. In this era, the Puget Mill Company also began using trucks to supply camps and bring some logs from the woods.

blocks were leveled, including the Pope & Talbot head office. Fortunately, the disaster spared the firm's main yard, where 8 million board feet were stockpiled. The mills sent an additional 54 million board feet down the coast. So much lumber flooded into San Francisco, however, that prices soon fell. Moreover, the latest national recession reduced overall demand between 1907 and 1910.

Beginning in 1907, western operators felt whipsawed by unstable national lumber prices. For the company substantial financial gain in 1906 turned into losses in 1907 and 1908. Wage cutting and mill suspensions produced a modest upswing in 1909 and 1910. Ports Gamble and Ludlow shut down for six months—far and away the longest such closures in the company's history to date. Despite a war-related pickup in lumber prices, only two years between 1911 and 1918 were profitable.

Small and overextended firms were badly hurt by the prevailing economic conditions. If bankruptcy was to be avoided or at least postponed, these operators literally had no choice but to engage in liquidation of timber, unrelenting production, and fierce selling tactics. Speculators endangered themselves acquiring high-priced timber and mill sites. Undercapitalized mills by the score were lured into precarious existence by the San Francisco disaster and the resultant brief lumber price inflation. In 1905, Washington became the nation's number-one lumber-producing state. More than five hundred mills turned out 4 billion feet of lumber, double the output of 1899. But by 1910 the state's 1,263 sawmills barely cut 3 billion board feet; huge unsold stockpiles accumulated.

The company participated fully in these various positive and negative trends. Port Ludlow was rebuilt in 1904. Night work commenced in the previous year at both Gamble and Ludlow, requiring three hundred additional employees. Port Ludlow management filled vacancies fifty at a time. Competitors paid higher wages, many of the recruits proved unfit for the task, and a high turnover rate became a common feature of the business. "We are short men," Walker complained, "and it is hard to keep supplied even with the things we have to keep running." Hiring by the shipload in Seattle, he secured men who, he lamented, had never worked before in their lives. Walker had no choice but to improve conditions and accommodations even for the misfits, as the era of crude mill quarters and bindle stiffs who carried their own rolled blankets was now history.

A sense of decline, if not impending doom, preoccupied older company leaders during these years. The diminished coastal trade, except in the year of the great earthquake, and the inroads made by the rail mills were depressing aspects of an uncertain age. The rail business in western lumber equaled and in some years exceeded the western lumber cargo trade. Profits never returned to nineteenth-century levels. New capital, new attitudes, and new methods of production and distribution dominated lumbering.

The company dropped, bit by bit, to secondary rank. Up "to twenty years ago," Ames commented in 1912, "the Puget Mill Company were the big people, today we are several points below the head of the list." Pope & Talbot had quite simply "not kept up to the times," said the Port Gamble boss. The Northern and Southern Pacific Railroads and Weyerhaeuser had become the super-rich of the industry, controlling by 1913 some 291 billion board feet of Northwest timber. Puget Mill, instead of ranking with these giants, was merely one among the sixty-one firms accorded status as large timber-holders.

DEVELOPMENT

AS RECESSION CUT DEEP, COMPANY EXECUTIVES RECONSIDERED THE LAND QUESTION AND debated how best to sell the firm. Rising timber taxes and insufficient dividends forced them to wring more than just lumber from the land. In 1906, Washington State assessed the firm $83,000 in timber taxes. The bill increased to $129,000 in 1907. The state and the counties taxed timberland on the value of the trees—which kept rising—rather than on the value of the land. Lumber interests argued that taxes should apply only when the trees were cut, in the same manner in which farmers were taxed for their crops.

George A. Pope made a special detailed study of the company's land holdings in the aftermath of the 1907 tax bill. Developing an expansive recommendation, the young executive recommended that the Puget Mill Company, its timber included, be sold in piecemeal fashion over a ten-year period. William H. Talbot argued, instead, that the corporation was worth more if sold in its entirety. In 1909 management set an asking price of between $23 million and $25 million. Pending a sale, the owners decided to meet cash and tax needs, and to pay more regular dividends, by selling sufficient land and timber each year to generate a $300,000 annual return.

Land dealings, but not actual property development, were prime components of the Pope & Talbot heritage. The company had always bought and sold undeveloped lots and acreage. The company owned most of the land between Seattle and Everett and significant acreage within the expanding city of Seattle. Some of these properties were worth more as real estate than as land used for growing more trees. Before 1900 the firm rid itself of 10,000 logged-off acres in western Washington. In that year it platted and sold the Washington

> *"My grandfather, George A. Pope, set about attempting to sell the company as soon as he was experienced enough to influence the Talbots and the Walkers. Both Cyrus Walker and William H. Talbot dragged their feet on any effort to find a buyer willing to pay a fair price, but as the two men aged, it was the younger Pope who would prevail. My grandfather, who with two childless sisters was the sole heir to Andrew Pope's interest in the partnership, wielded more influence as one individual than any of the Walker or Talbot heirs. The business was sold to Charles McCormick, despite Walker's and Talbot's reluctance and despite my own father's opinion that McCormick was a gambler. My grandfather got his way, but ironically, it was the Great Depression that would return Pope & Talbot to the families and their future generations. McCormick was hard hit by the Depression, and although he cut down all the trees trying to pay his debts, it wasn't enough. In the meantime, my father had married Captain Talbot's great-granddaughter, Harriet Brownell, and had begun to think about the future. This marriage left my grandfather and the Pope side of the business in effective control of the company. My grandfather, in turn, left my father in control. In addition, the Talbots had not developed an acceptable leader. From the late 1920s on, Pope & Talbot would be led by Popes."*
>
> —Peter T. Pope

Park neighborhood in Seattle, with Ames included among the mansion-building purchasers. Seattle's growth meant that considerable cutover acreage might be profitably disposed of in the form of residential subdivisions.

The 1909 decision to generate annual returns from the land produced, in addition to property development schemes, a decision to directly undertake logging. In 1910 the Puget Mill Company opened its first logging camps, designated Camps One and Two, on Gamble and Ludlow Bays. Although Talbot criticized destructive "highballing" methods (working at such a high rate of speed and leaving little growing), he had no trouble following the conventions of industrial timber removal. The company cut both the best trees and the supposedly inferior trees the first time over a given tract. Logging took place "only once," according to Ames, "and when you get thru there is from 3 feet to 5 feet of litter and debris all over the land; the only trees standing are a stub here and there . . . and the spar tree."

Far distant from the mills, management directed logging activities on company acreage adjacent to Seattle's expanding city limits. Between 1911 and 1920, 3.5 million board feet of virgin cedar, fir, spruce, and hemlock were cleared north of Seattle and south of Everett. Initially promoted as stump farms for would-be agricultural settlers, a 7,000-acre cutover opened in 1917 as Alderwood Manor, the firm's first centrally located real estate development.

Plans to sell the entire company failed, but not for want of trying. In 1912 most of the holdings were briefly optioned to a Spokane banker attempting, futilely as it turned out, to combine cargo mills and timber in a trust arrangement. In the meantime Ames proposed major policy changes to restore profitability. Absent interest among the younger family members, he suggested that in the future the company ought to be run by professional managers. A serious buyer not appearing, however, several of the Ames proposals were adopted.

Ames advocated change within both the company and the industry. Since 1901 the firm had helped pioneer the trade association movement. Mills on Puget Sound and Grays Harbor formed the Pacific Coast Lumber Manufacturers Association (PCLMA). Combining American and Canadian operators, the West Coast Lumbermen's Association succeeded the PCLMA in 1911. Both associations labored to counter adverse market conditions and the unions. They lobbied for federal regulation of railroad rates and lower state taxes. They worked for improved state and private fire protection. They urged the adoption of new forestry doctrines to end heedless logging and unprofitable destruction of timber. In sum, they worked for limited reform, for changes that would reduce savings for the bigger operators, and against mandatory federal regulation of the industry.

Price manipulation was the major goal. Until 1906, when federal antitrust authorities began an investigation, members openly rigged prices, or at least attempted to do so. By 1913 the continuing Justice Department inquiry made Talbot and Ames fearful. Ames hid association price lists from all prying eyes. Historically, price-fixing had worked best during good times and had failed during hard times. The industry's habitual inability to cooperate sooner or later doomed such schemes.

Ames, and through him the firm, dominated association councils for nearly a quarter century. He presided over the Pacific Lumber Inspection Bureau, which devised and administered uniform grading standards, from 1906 until 1930. The bureau quickly won the respect of both buyers and sellers, primarily because Ames successfully withstood pressure to certify inferior shipments. In 1911 he played an instrumental role in creating in Washington one of the country's first state workmen's compensation systems. The per capita rate of death and disability in the logging and milling sectors was historically five times greater than in any other industry. Thinking had changed since the days of Cyrus Walker. Back then, except for provision of immediate medical treatment, the venerable Boss Logger opposed any and all forms of assistance to injured employees. According to that traditional doctrine, accidents were caused by carelessness and therefore the victim's responsibility. Company leaders now agreed, however, that workmen's compensation was a sound business practice, reducing discontent, lowering costs, and eliminating the need to deal with private lawsuits.

TURMOIL

NORTHWEST LUMBERING RODE A LABOR ROLLER COASTER FROM 1906, WHEN THE Industrial Workers of the World (IWW) was organized, through the ups and downs of the First World War and into the turbulence of the early 1920s. In mill towns and in urban centers the IWW (the "Wobblies") struck for higher wages and the eight-hour day in 1906, 1907, and 1913. Company executives literally hated the militant union. Ames hired detectives to infiltrate the IWW but failed largely in the attempt to root them out of the woods and the mills.

The First World War, begun in August 1914, sent the foreign lumber market into a temporary tailspin. European demand evaporated, prices collapsed, and mill failures reached their highest point in a half century. In a stab at depression-induced efficiency, Pope & Talbot dissolved five of its twenty affiliated companies. The war, however, soon inaugurated an economic boom. Orders poured in from the war zones. Domestic demand quickened. Almost overnight, wooden shipbuilding attained major significance as an adjunct to the Northwest lumber industry.

FIGHTING THE WOBBLIES

The conditions at a company supplier, English No. 2 Logging Camp, photographed by company timber surveyor L. Heath in 1898, helped bring on strikes by Wooblies before World War I.

FOR SEVERAL YEARS THE COMPANY FELT hammered by the Industrial Workers of the World (IWW). The IWW pledged a workers' utopia and the end of capitalism. From its formation in 1906 it led dramatic work stoppages across the continent. In the Northwest it tapped into discontent over extended workdays and the treatment of loggers and mill workers. In mill towns and urban centers in 1906, 1907, and 1913, the IWW struck for higher wages and the eight-hour day. Countless job-hungry Northwesterners helped drive them back to work.

Wobblies advocated industrial sabotage but were more often the victims of violence themselves. In the Pacific Northwest employers, most American Federation of Labor (AFL) unions, local vigilantes, and government authorities targeted them. Detectives hired by the company's Edwin G. Ames and trade associations infiltrated the IWW and partially rooted it from the woods and mills.

The First World War changed the production and labor scenes. After American lumbering's initial near collapse at the outbreak of fighting in 1914, northwestern mills grew sleek with orders. The country went to war in April 1917. In July, strikes by the IWW and an allied AFL timber union closed most mills and camps in western Oregon and western Washington. Company leaders felt profoundly challenged by the IWW. Ames now joined trade association opponents of the eight-hour day. The Lumbermen's Protective Association, headed by E. S. Grammer from a Puget Mill Company logging subsidiary, formed to stiffen the operators and stop the Wobblies. Hardliners like Ames privately recognized that the shorter day was inevitable but insisted it appear to be a voluntary gesture after an IWW defeat.

The company needed detectives and sheriff deputies to keep Port Gamble and Port Ludlow open. Chaos spread once the IWW returned to work in September 1917 and slowed production. Enraged employers reluctantly agreed to federal intervention. They traded the eight-hour day—and more liberal conditions if they could not be resisted—for a government crackdown. The Loyal Legion of Loggers and Lumbermen (4-L), combining employers and employees into a coalition of company unions, furnished workers. The army's Spruce Production Division assigned soldiers to guard and work in the plants. The army helped ward off an AFL organizing drive among loggers, and the IWW was driven into sullen submission or into nearby towns.

But Ames wanted old-time prerogatives reinstated. He half-heartedly cooperated with the pro-eight-hour 4-L day after the war ended. His mills escaped the AFL strikes and the other labor turmoil of that time. In 1922 he arranged for sixty replacements to work a reinstituted ten-hour day. Eight-hour shifts were restored about a year later, after the IWW made inroads in company logging camps. Labor peace settled on the camps and mills until the early 1930s.

In 1916, Congress lifted antitrust barriers to exploitation of overseas trade. The resultant Douglas Fir Exploitation & Export Company, combining thirty-eight cargo mills into a joint exporting agency, locked in customers, fixed prices, spread orders among members, and provided shipping. Widely trusted in the industry, William H. Talbot served as president of the new commercial venture.

A Wobbly resurgence in 1917 frightened the otherwise happy operators. In July the antiwar IWW and a loosely allied American Federation of Labor (AFL) timber union closed most of the mills in western Washington and Oregon. The national interest was endangered, conservatives and other patriots convinced themselves. Believing the strike to be nothing less than a treasonable wartime rebellion, Ames took the lead in forming the Lumbermen's Protective Association to fight the Wobblies.

Detectives and county deputies prevented serious loss of production at Port Gamble and Port Ludlow. A wage increase was instituted, although the benefits were partially offset by higher charges for room and board. Returning to work in September, the IWW instituted a deliberate slowdown campaign, following all safety rules and other regulations to the fullest extent. Newly enraged, employers called for federal intervention, offering the eight-hour day in return. The army responded, forming the Spruce Production Division, which effectively militarized the forest for the duration of the war, and a broadbased pro-company union, the Loyal Legion of Loggers and Lumbermen.

At Ludlow and Gamble the IWW was quickly overwhelmed, and high rates of production were secured and sustained. To ensure keeping the mills open, Ames went along with army-mandated improvements to living and working conditions. Upon the return of peace, he thought, traditional managerial prerogatives could be restored. In 1919, however, Seattle experienced a general strike, the AFL closed mills in Bellingham and on Grays Harbor, and a deadly riot erupted in Centralia. Reluctantly, Ames continued to support the eight-hour day and the antiradical Loyal Legion. In 1922 the company finally brought in replacements willing to work a reinstituted ten-hour day. The resumption of Wobbly activity, however, soon forced the restoration of eight-hour shifts.

WINDING UP

A DREADED REFRAIN BEAT ON POPE & TALBOT IN THE IMMEDIATE POSTWAR PERIOD: the new federal income tax, the federally mandated refiguring for taxation purposes of the book value of the firm's timber, and the prospect of heavy inheritance taxes should William H. Talbot or George A. Pope succumb. Talbot was now in his sixties, in failing health, and depressed over his brother Frederick's death in 1919. The company limped forward into a period of increasing uncertainty.

National economic developments spun the company in unhappy directions. Lumber demand and prices spurted in the peacetime spring of 1919, but recession, settling briefly on the country in 1921, wiped out the gains. A miniboom followed in 1922 and 1923, succeeded by an unexpected market slump in 1924. Overall Washington production remained at a high level, with prices falling sharply. Everything, Ames lamented, was going down and down. By 1925 lumbering ranked with the coal, rail, and textile industries and with agriculture as a distressed economic activity.

The on-again, off-again effort to sell the firm proceeded in fitful fashion. A competing lumber company cruised the timber in 1919 but decided not to submit an offer. Another potential buyer dropped out after

Broadmoor, the company's upscale residential development in Seattle, opened in 1924. Many amenities, such as the golf course, pictured here in the 1960s, helped maintain its longtime value as a desirable place to live.

failing to secure financing. After submitting an attractive bid, a shipbuilding firm killed the prospective deal by arbitrarily reducing the amount of the down payment. "We have entered a campaign to make our business profitable," Ames wrote as 1920 opened and new suitors failed to appear. The goal was truly formidable. The firm competed, Ames realized, in a largely revolutionized international business climate. The newly important lumber brokerages had drastically altered the methods of the waterborne trade. The profitability of cargo mills trading offshore depended, to a considerable degree, on the rates and vessels assigned by brokers.

A big East Coast market, meanwhile, dictated that far more output be channeled there than before the war. By 1923, as western mills shipped increasing amounts of lumber through the Panama Canal, Ames considered the New York trade a godsend. But the firm's outmoded facilities and dependence on ships belonging to other concerns prevented full development of the potential Atlantic coast business. At the same time, the lack of rail connections kept Pope & Talbot out of the Midwest. The Florida real estate boom did little for sales, since southern mills were much closer to consumers.

If the company intended to truly compete on a national scale, it had to rely on its low-cost timber to overcome its inefficient, high-cost mills and steep freight rates. From 1921 to 1925 the Port Gamble mill, a relic built in 1870, ran heavily in the red. Port Ludlow, rebuilt in more recent times, performed only a bit better. In 1924, Talbot concluded that further operation of the Port Gamble plant was suicidal and ordered its closure. The continuing speculative scramble in the forest represented the one positive note, and Ames celebrated "the rapid increase in the actual value of [our] growing timber."

Most of Pope & Talbot's postwar lumber, according to company historians Edwin T. Coman, Jr. and Helen M. Gibbs, "sold at a loss, for [its] old mills on the Sound simply could not produce lumber to be sold at competitive prices." Milling costs in 1922 were almost a dollar per 1,000 feet above the industry average.

"The higher cost of manufacturing," explained Ames, "is due to storage of lumber in the yard for air drying, kiln drying, and the resawing of lumber into other sizes, the careful grading of lumber, and the manufacture of lumber into finished stock like shiplap, ceiling, rustic and flooring." The veteran lumberman uttered a classic lament: "The more you do to it the more it costs."

There were some bright spots in the gloomy canvas. During the early 1920s Japan became an important market for mills on Puget Sound and Grays Harbor. The nation imported a yearly increasing tonnage, particularly in the form of so-called Jap Squares, large, roughly squared timber. Meanwhile, the Santa Barbara, California, earthquake of 1924 brought short-term profits. Longer-range profit seemed possible in Seattle, an outlet once largely ignored. Shifting the central sales office from San Francisco to Seattle, veteran manager Richard Condon began tapping both local and rail-connected markets.

George A. Pope Sr. (1864–1942)

The company also ventured in 1924 into a new, upscale land development in Seattle. On a logged-off Lake Washington parcel of 215 acres, it crafted in Broadmoor a "Country Club within a City." The firm laid out handsome streets, parks, the region's first complete underground utility network, and the latest American status symbol—a golf course. The first sixty-five substantial houses were soon under construction.

Another would-be buyer for Pope & Talbot surfaced that year, a partner in one of the brokerage firms that had so fundamentally altered the lumber business. Charles R. McCormick would, according to one contemporary associate, "buy anything." Some family members considered him a wild speculator and an entirely inappropriate suitor. Despite that, on July 25, 1925, following intricate negotiations, the families sold their principal holdings to the McCormick Lumber Company. The sellers thereby gave up on the traditional goal of maintaining family control of Pope & Talbot. There would be no succession to leadership beyond the second generation of Popes and Talbots. Or so things appeared.

CHAPTER FOUR

TRAVAIL

1925-1940

MCCORMICK

CHARLES R. MCCORMICK AND SIDNEY M. HAUPTMAN WERE MIDWESTERN NATIVES transformed by character and circumstance into high-flying Californians. Starting out with an office and a telephone, the partners eventually developed one of the four largest lumber brokerages in the West. They also expanded into mining and California and Mexican railroading. When not preoccupied with these endeavors, they became lumber wholesalers and shipowners. McCormick and Hauptman led the way in fundamentally changing the methods under which traditional cargo mills did business.

On the brink of the First World War, a dozen McCormick steam schooners served West Coast lumber ports. Unlike Pope & Talbot the McCormick group had no prejudice against steam transportation. Eventually, it bought modern steel coastal vessels, including first-class passenger ships, and scheduled regular calls at an ever-expanding number of port cities. Cargoes were guaranteed in part via creation of a sawmill complex at Saint Helens, Oregon. At sixty sailings a month McCormick was by the mid-1920s the largest coastal lumber carrier. Almost three-quarters of the Northwest lumber sent by sea to California in that decade was carried aboard McCormick vessels.

The McCormick Steamship Company was officially organized in 1920. The federal government provided key assistance through subsidies, mail contracts, and low asking prices for war surplus steamships. McCormick formed joint operating agreements with some lines and absorbed others. The Munson-McCormick service, for instance, inaugurated a weekly run in 1923 between the Atlantic and Pacific coasts. McCormick subsidiaries maintained, loaded, and unloaded vessels on both coasts.

McCormick had trouble finding enough northbound cargoes on the Pacific coast run. A similar shortage often precluded efficient use of the big steamers sailing from eastern ports to California. The Munson-McCormick service helped lumber producers, the Saint Helens mill included, develop an eastern market for Douglas fir. Once in charge of the Puget Mill Company, McCormick forged an even closer relationship between the cargo mills and eastern customers.

McCormick regularized mill trade with Latin America. In 1926 the Munson firm and McCormick jointly organized the Pacific Argentine Brazil Line (PAB), later a wholly owned McCormick subsidiary. Refrigerated vessels and passenger craft called monthly along the eastern coast of South America and at

(right) McCormick coastal passenger and cargo vessels alongside Portland's McCormick docks on the Willamette River. During the 1920s McCormick had more vessels on the Columbia and Willamette Rivers than any other single company. *(Harold M. Brown, photographer)*

(below) Float of McCormick S. S. Company in parade of May 2, 1925.

Cuba and Puerto Rico. Western rice, lumber, paper, and other manufactured items flowed southward to Cape Horn. The direct service was instrumental in California, becoming a major processor and distributor of coffee, chocolate, wax, nuts, spices, drugs, and fertilizer. PAB tonnage growth was fairly steady, at least until the Great Depression. At one point the firm handled 45 percent of the commerce on the route.

The McCormick group believed that it could make, sell, and transport anything related to wood. Easterners and Californians alike invested in the shipping and manufacturing enterprises. The Saint Helens mill, originally opened in 1909, spawned dependencies. Moving opportunistically whenever and wherever profit appeared obtainable, the group piled debt on debt. The acquisition of Pope & Talbot was merely the biggest and most highly leveraged of many similar deals. Stock market and banker optimism on the one hand and seller pessimism on the other hand drove the transaction to completion.

A key McCormick move in March 1925 made the final arrangement a reality. The group merged seven Oregon and California firms, forming the Charles R. McCormick Lumber Company of Delaware. The combined assets, just under $5 million, creatively emerged from the arrangement as more than $14 million. In July, McCormick Lumber took over the Pope, Talbot, and Walker interests for $15 million. A 10 percent down payment was made—taken by the selling families in stock—with the remainder to be paid over fifteen years, at 5 percent interest. The purchaser agreed to replace the Port Gamble mill, expending on the project $1 million of the $2.5 million in corporate stock separately acquired by the Puget Mill and Rainier Investment Companies. Only Hauptman thought the transaction financially unsound.

The families financed their own buyout, but not of everything and not without safeguards. The Puget Mill and Rainier Investment Companies held mortgages on all McCormick Lumber properties, camps, and timberland. These covered the McCormick manufacturing center at Saint Helens, retail lumberyards in southern California, and sales offices in New York and elsewhere.

Legally, the San Francisco–based McCormick Steamship and McCormick Intercoastal Steamship Companies, along with their subsidiaries and affiliates, were not part of the transaction. McCormick's substantial stake in these moneymakers nonetheless made the deal practicable from a financial point of view. None of the lines carried a heavy debt burden that might reduce net earnings. McCormick ran the largest

Charles R. McCormick (1871–1955).

steamer service on the coastwise run. The three-year-old intercoastal service was also making a mark. The lines had, or soon opened, terminals in Los Angeles, Portland, San Diego, San Francisco, Seattle, Tacoma, and several eastern ports. The fleet gave the old cargo mills a new lease on life. Pope & Talbot executives believed that profits from shipping would surely help pay the mortgages.

McCormick Lumber obtained a Port Gamble mill badly in need of replacement, one at Port Ludlow needing an upgrade, a few affiliates, a small sales unit, and what it most dearly needed—92,254 timbered acres with an estimated 3 billion board feet of timber. Overcutting by the Saint Helens sawmill had nearly exhausted that plant's 4,000-acre wood supply.

The former owners consolidated or closed their tug and logging companies. In 1930 the Talbots and George A. Pope Sr., who had purchased an interest in the firm in 1927, sold the sole Pope & Talbot asset, the San Francisco lumberyard, to McCormick Lumber for $225,000. According to one of the city's newspapers, this deal severed "the last connection between retail lumbering here and the two families, leaders in finance, real estate, shipping, and, naturally, society." Actually, the sellers had rid themselves of fewer than half their total array of corporate holdings. The Puget Mill and Rainier Investment Companies still held 47,051 acres of timberland, plus cutover land and real estate in Seattle and its vicinity. Both family firms turned to aggressive developing and selling of timber and real estate. Individual family members also retained California ranches and general real estate, stocks, and bonds. They were not yet out of business.

EXPANSION

THE MCCORMICK LUMBER COMPANY RAPIDLY MODERNIZED AND EXPANDED ITS NEW properties. The firm built Port Gamble anew, improved Port Ludlow, extended logging operations, and generally integrated all manufacturing and shipping functions. Mill output increased to 300 million board feet a year, despite the worsening lumber recession. National lumber production dropped from a postwar high of 41 billion board feet in 1925 to 36 billion in 1928 and to 26 billion in 1930. Thanks in part to productivity gains, McCormick, Weyerhaeuser, and some of the other big producers bucked this downward trend. Betting on domestic construction and consumer spending and in some cases saddled with debt, they raised overall western output, contributing to the industry's general and chronic overproduction.

The improvements cost far more than expected. The final outlay for the new Port Gamble mill was $2 million, a far cry from the anticipated $400,000. No manner of equipment seemed too expensive to install and run. The well-known logger Alex Polson compared McCormick's spending to that of a drunken sailor. The investments did at least bring short-term increases in production and employment. In 1927 about 1,000 people worked at Gamble, 450 at Ludlow, 800 at Saint Helens, and 500 in the logging camps. Four years earlier Port Gamble had employed barely a hundred workers.

Longtime employees later faulted Charles McCormick's stewardship. They charged that he paid little attention to spending and administration, built and modified mills on too large a scale, and ran the facilities in an extravagant and wasteful fashion. Operating costs mounted in a completely unchecked manner. Supervisors ignored advice from veteran employees and hired outside experts. Integration of new and old management proved to be a difficult challenge. The critics conceded, though, that Port Gamble would not have survived without the modern plant and equipment.

A 90-foot Lidgerwood steel tower near Port Ludlow supplements Camp Talbot's high-lead logging in 1912. The first steel spar skidders, Lidgerwoods, moved under their own power on rail, using blocks, drums, and cables to yard and load logs. Lidgerwoods replaced tree-rigged spars and high climbers, and, when oil-fueled, saved valuable timber from the steam engines.

In the 1890s the company installed heated kilns at Port Gamble and Port Ludlow to improve lumber quality. Here, at Saint Helens in the 1920s, McCormick renewed these drying kilns and expanded the air-drying facilities.
(John D. Cress, photographer)

THE BIG MILL AT SAINT HELENS, OREGON

IN 1909, CHARLES R. MCCORMICK AND HIS BROTHER Hamlin McCormick, ambitious California lumber brokers, ship owners, and entrepreneurs, arrived downriver from Portland on the lower Columbia River. In tiny Saint Helens they opened a modern sawmill. Upward to four steamers from the McCormick and other lines could load at the dock of what local people called the Big Mill.

Over the next twenty years the McCormicks opened more firms nearby. They set up a plant to apply creosote, a preservative, to railroad cross ties, piling, structural timbers, and other rough products. They helped incorporate the Saint Helens Pulp and Paper Company, which bought the Big Mill's sawdust and wood chips. They opened a broom handle plant to use its slabs, and the Fir-Text Insulating Board Company, another sawmill waste buyer. They opened but closed two other sawmills and a wooden schooner shipyard. The McCormicks became the area's biggest employer and reputedly paid top dollar in the low-wage lumber industry. Local manager Hamlin McCormick was well liked.

At the opening of the century, Oregon ranked seventeenth as a lumber state to number-one Washington. It moved up rapidly as the McCormicks and others acquired forests and opened mills. The Pope & Talbot lumberyard and their families' California lumberyard chains felt serious pressure from the Oregon rail mills. By 1930 the state's sawmills outstripped Washington production two to one. Many Oregon mills by then were new, and a few were very big. Numbers of them had or would close once they had cut all their own timber. That was nearly the story of the McCormicks in Saint Helens.

From their beginning in California, the McCormicks looked beyond the West. They integrated their Oregon and California wood product and shipping firms and marketed on almost a hemispheric scale. They sent a tide of Saint Helens

wood products into the national, Caribbean, and South American markets.

After acquiring Pope & Talbot, the Charles R. McCormick Lumber Company, like many competitors, rapidly cut timber. It nearly exhausted the Big Mill's 4,000-acre tract. Only 334 million board feet of timber remained in its cutting circle in 1940. The McCormicks meanwhile cranked up the Big Mill from 50 million board feet in 1925 to an unheard-of 113 million in 1928. The Depression drove down production to around 25 million board feet in 1931. McCormick Lumber's sophisticated sales-distribution network cut prices and dumped output.

Its shipments wavered between 42 and 76 million board feet between 1932 and 1937 and had fallen to 38 million when the new Pope & Talbot Lumber Company took over McCormick properties in 1938. The outbreak of World War II catapulted the Big Mill's shipments to over 111 million board feet in 1939 and inspired streamlining manufacturing processes, enlarging mill capacity, and expanding its cutting rights in private and federal forests. War had brought rail to Saint Helens, and most production left that way. The Big Mill's cargo mill roll shrank.

Finding, training, and retaining enough workers plagued the round-the-clock wartime plant. Many employees knew nothing about mill work and required housing. The transients "gave us the most grief," the company admitted. For the first time almost three dozen women temporarily received blue-collar jobs in the mill and creosoting departments.

The Big Mill claimed Oregon's highest wartime sawmill output. Shipments of lumber, pilings, staves, and ties climbed between mid-1941 and mid-1944 from more than 139 million board feet to past 196 million and declined to more than 174 million as the war ended in 1945. Many orders were rushed. Some lumber traveled to secret locations, including 35 million board feet to build the atomic bomb plants in Hanford and Richland, Washington. Proud of its prodigious wartime record, company executives now had to find enough timber to keep the Big Mill operating satisfactorily as postwar lumber demand skyrocketed.

The Saint Helens mill spreads along a Columbia River slough, about 1939. Pope & Talbot just missed taking advantage of the heavy federal wartime investment in big mill facilities: the Saint Helens expansion and upgrading finished about the time war broke out in Europe.

Missteps certainly occurred. At one logging operation, for example, a railroad bridge was rebuilt after a spring flood then abruptly demolished when a decision was made to shift to truck hauling. The structure was soon reconstructed in order to recover locomotives left on the wrong side of the stream. Despite promises, McCormick did not retain all of the old employees. Nobody had a pension, so Ames requested that the traditional policy of providing aged workers with undemanding tasks be continued. Instead, mill management fired all employees unable to keep up with the work pace.

The new owners replaced Port Gamble's nineteenth-century plant with an electrically driven facility. The mill was far bigger than originally contemplated. New saws, pony saws and resaws, extensive planers, automatic loaders and sorters, and a drying shed were deployed. Motorized Ross carriers simplified movement of lumber. Colby cranes lifted railcars on the barges for transportation to Seattle. Hog fuel bunkers replaced the old polluting burner. Slabs, sawdust, and sweepings were sold or burned to generate power.

Less was done at Saint Helens and less still at Port Ludlow. Both mills renewed their docks and installed cranes and Ross carriers. Saint Helens dedicated $100,000 from the sale of its electric plant to modification of facilities. In addition to remodeling the Big Mill's drying kilns, management installed a new band saw, which used a toothed continuous blade, air-drying sheds, and mechanized handling and loading equipment. The mill's capacity increased to 375,000 board feet a day. Shipments from Saint Helens were accelerated to counter falling prices and pay debt: from 50 million board feet in 1925 to 99 million in 1927 and 133 million in 1928.

A tide of lumber, spars, masts, piling, bridge timber, railroad ties, poles, shingle bolts, Jap Squares, and pulp wood flowed from the three McCormick mills. The Saint Helens wood preservation plant treated piling and poles with creosote. Nearly all of the low-grade material went to Japan or the western United States. The demand for Jap Squares, mine timber, and rail ties appeared to be inexhaustible. McCormick also dealt in high-value products. Remanufacturing facilities at Port Ludlow and in California produced framed and planed lumber.

The firm wholesaled and retailed lumber in the West, particularly in California. The southern California population boom between 1925 and 1929 cast a bright light on gloomy times. New homes, businesses, and factories used McCormick lumber. The seventeen West Coast paper mills opened after 1925, together with existing manufacturers, purchased low-cost pulpwood and sawdust. The corporation even challenged the southern mills. McCormick sold lumber in Florida and was the largest supplier of Douglas fir timber and ties to consumers in Lake Charles, Louisiana. In Hawai'i, however, an important historical customer, Lewers & Cooke, was lost to price-cutting rivals.

McCormick Lumber invested heavily in logging. Ending the contractor system, management turned the former Puget Mill Company camp across Gamble Bay into Camp Gamble. At great cost Camp Cowlitz was opened on the distant Cowlitz River. Saint Helens received logs from Kelso, on the Washington side of the Columbia River. Other new operations included Camp Talbot, outside Port Townsend, and a logging setup near Castle Rock.

The trade papers portrayed pleasant woods communities of professional foresters, craftspersons, supervisors, laborers, service personnel, and actual loggers. Gravel streets, lawns, flowers, post offices, schools, stores, fire protection, and modern utilities graced the scene. Families lived in comfortable cottages. Tidy bunkhouses accommodated single people. Residents owned automobiles, modern highways having put the camps within convenient reach of the outside world.

In the mid 1920s, McCormick upgraded its mills' mechanization, contributing to the sawmill and forest productivity drive of that decade. At Port Gamble *(pictured)* and elsewhere, Brownhoist cranes and Ross carriers replaced animal power and reduced human handling of lumber. *(John D. Cress, photographer)*

Even when compared with the mill investments, the sums spent on the camps and logging equipment were extravagant. Camp Talbot ran 160 trucks between yarding sites and log booms in Quilcene Bay. Cowlitz had two fifty-ton locomotives and strings of rail log carriers. The corporation employed steam yarders and mechanical skidders with "compound engines," increasing productivity while lowering labor costs. Four new Lidgerwood steel skidders, costing $60,000 apiece, were instrumental in harvesting an average of 200,000 feet a day per site. Highballing in the woods created a need for additional roads, bridges, culverts, and rail lines. By 1934 seventy-five miles of McCormick logging railroad snaked through company timber holdings.

Logging went on six days a week, regardless of all but the most severe weather conditions. McCormick compelled Camps Cowlitz, Gamble, and Talbot to increase their total cut from 850,000 board feet to more than a million per day. Loggers highballed through age-old fir and mixed stands of fir, cedar, spruce, and hemlock. Only output counted. Trees fell at a pace that would have horrified the old Puget Mill Company. Once-husbanded resources, however, guaranteed the continuation of business at Port Gamble and prevented closure of the Saint Helens mill.

By 1930 two-thirds of the timber acquired in 1925 was gone. Old-time loggers later recalled the incredible amount of waste left behind in the woods. McCormick Lumber deserved prosecution for overlogging, one company veteran fumed. On a national basis "rapacious cutting reached its apogee during the 1920s and early 1930s," historical geographer Michael Williams writes. "The forest was viewed as a resource that could be exploited rapidly in order to pay for the heavy capital investment needed to deal with the large trees of the West." And in McCormick's case, as well, to keep the mills and ships in business.

CONTRACTION

RED INK RAN FROM THE MCCORMICK LUMBER COMPANY BOOKS. THE CORPORATION, unfortunately, had commenced work at the outset of a lumbering industry recession. After a year of modernization and expansion expenditure, McCormick teetered on bankruptcy. No Washington or Oregon sawmill earned a profit in 1926, the annual report stated in self-defense. Hearing the bad financial news, Edwin G. Ames was unable to sleep at night. Scheduled payments to the former owners fell delinquent. Two of the big McCormick steamers were transferred to the Puget Mill Company in partial compensation.

Cash-poor executives adopted the conventional responses—cutting costs, dumping output on the market, and improving efficiency where possible. The Saint Helens mill sold sawdust to a nearby pulp and paper mill. McCormick issued $4 million in bonds and, in return for further mortgages, obtained additional Puget Mill Company financial assistance. Management held losses to a moderate level in 1927 and actually realized a modest profit in 1928.

By this time western mill operators had already turned to another traditional remedy—the merger. In

Piles protectively treated by the Saint Helens mill went into this 86-foot-high railroad bridge, under construction when this was taken in Oregon in 1932. The company did a lively business in treated piles for bridges, docks, warehouses, and other big structures.

A McCormick ship carries 8,885 pieces of 14-inch mining butts and almost 1.8 million board feet of treated Douglas fir timbers, 1928. Lacking passenger vessel glamour, such workhorses were a common site in U.S. waters.

1926 bankers began working on a prospective merger of McCormick and forty other Pacific Northwest concerns. The plan failed. In the face of a decline in construction, meanwhile, the invaluable Atlantic coast market lost momentum. Unable to reduce output, Charles McCormick dumped wood products in the East. At the risk of considerable personal notoriety, he sold consignments by the shipload at almost any price offered. If this practice was continued, longtime associate Sidney Hauptman warned, the firm would surely fail.

Facing further delinquencies, William H. Talbot and George A. Pope forced management changes in April 1929. Hauptman became president and McCormick, board chairperson. Two Puget Mill Company representatives joined the board, the former owners assuming control of the executive committee. Ordered to drastically cut costs, Hauptman reduced the net loss for 1929 to $3,000. The next year, however, McCormick lost more than $800,000. Hauptman sold three California retail lumberyards and abandoned the San Diego wharf.

The firm avoided bankruptcy in June 1930 when the Puget Mill and Rainier Investment Companies canceled $2.5 million in principal and $183,530 in delinquent interest and waived further interest payments. In exchange, McCormick Lumber bought out the minority owners of the McCormick Steamship Company and pledged that firm's future earnings to the Popes, the Talbots, and the Walkers.

The stock market collapse in the fall of 1929, followed by the Great Depression, set off long-running crises. Pacific Northwest lumber output plummeted between 1929 and 1932, the Washington figure dropping

ALDERWOOD MANOR

ALDERWOOD MANOR OCCUPIED ABOUT 7,000 ACRES of company stumpland halfway between Everett and Seattle. It opened in 1917 as the Puget Mill Company's first coordinated real estate development. "Stump ranch" lots sold for $200 to $400 an acre. Buyers obtained five- or ten-acre or larger plots and backing. The company had seldom acted as a property developer, and when it did, it tended to be in high-income urban areas rather than in stumpland sales to farmers. The company had successfully platted and sold the Washington Park neighborhood to Seattle mansion builders, including its own Edwin G. Ames, and 80 acres north of the University of Washington. In 1924 the Puget Mill Company crafted the handsome Broadmoor subdivision within Seattle.

Big twentieth-century timber companies added land sales divisions. Before the 1940s they believed that settlement, not private (and certainly not government) reforestation, best salvaged their cutover land. Left unused, this growing acreage appeared valueless, and, because property taxes continued, an income drain. Lumber firms in Michigan, Minnesota, the Pacific Northwest, and Wisconsin grew active in the "back-to-land" movement. Similarly, arid-land sellers promoted irrigation to back-to-the-landers.

These sellers shared a widespread belief that the poor were anxious to flee crowded cities for small farms in underpopulated areas. Veteran timber surveyor Fred Drew estimated that annually the Puget Mill Company cleared trees from 6,000 acres. On August 17, 1909, he told George A. Pope that "as many land hungry Europeans are coming to this country, I think the demand for" stump farms will increase. Between 1900 and 1920, Washington added almost seventeen thousand farms; small ones in cutovers bulked largest.

Many city people without financial resources did want to become independent by owning a farm and building it with family labor. Offering 10 percent down payments, liberal credit, and other aid, the company conducted high-pressure Alderwood Manor campaigns among poor, foreign-born city folk. The Puget Mill Company signed over more than $2 million in contracts in the Chicago area alone. It built roads and cleared an acre in each 40-acre plat. If paid, it constructed modest homes and outbuildings. Managers erected a hotel for prospective buyers, schools, and a clubhouse and library. They promoted mutual help, hard work, wholesome diversion, town meetings, patriotism, and snug futures as small-time berry farmers or dairy or poultry raisers. For a quarter-million dollars, the firm established a Demonstration College and 30-acre farm directed by an expert. It drilled wells, ran power lines, and advanced farm-improvement costs.

A couple pose with several days' worth of eggs to promote the company's Alderwood Manor, 1920s. An expert adviser and an egg producer cooperative helped secure Alderwood Manor a good market before more efficient Washington poultry raisers made the venture unprofitable. (Pierson & Co., photographer)

(above left) A modest Alderwood Manor farm home, garage, water tower, and poultry house, 1920s. Snags and stumps adorn the adjacent fields. *(Pierson & Co., photographer)*

(above) Men in business suits head through the parking lot of Alderwood Manor's main store, 1920s. Little other than electrical lines and snags surround the area. *(Pierson & Co., photographer)*

According to company promotional material, Alderwood Manor stood "For the Individual, a Little Land and Liberty; For the Community, Co-operation and Efficiency." Its first principle was "Feed yourself." Filbert-raising was tried and abandoned. Poultry-raising was then heavily encouraged as the best way to farm a generally desolate waste of stumps, snags, and water-soaked logs. After its egg supply overwhelmed local markets, the resident expert helped organize a producer cooperative that sold millions of cases in New York.

World War I's prosperity and Seattle's growth made sales and real estate profits thrive. One year Alderwood Manor remitted more than $1 million to the parent firm. But in the 1920s it began running deficits. Almost fifteen hundred people were raising poultry and eggs on the stump farms when an agricultural recession struck in 1922. Three years later Alderwood Manor was in serious financial trouble. Consulting irate, hard-pressed residents, company executives exposed poor business and financial practices. New managers were hired, costs cut, and the demonstration farm closed. William H. Talbot lost his enthusiasm for a colony of independent small farmers. It was not included in the company sale to McCormick Lumber Company. Occasional future profits were only on paper.

Making these farms profitable for residents was probably impossible. Despite assistance, colonists lacked the experience and capital to compete with better-situated lowland Washington farmers who turned to poultry-raising during the 1920s. The agricultural depression of the 1920s and the Great Depression caused suffering in the nation's cutovers. Without extensive private loans or government aid, the colonists probably could not have succeeded. Most adults began commuting to jobs forty to forty-five minutes away in Seattle. One couple became bootleggers.

During the Depression, Dust Bowl refugees and unemployed city dwellers settled on western Washington's cutover lands but were not permitted to settle in Alderwood Manor. Any resident paying property taxes was permitted to remain without further payment until 1935, when 105 properties were taken back, rehabilitated, and resold, some at a loss. Ultimately, about $2 million of once-inhabited land returned to the company. In 1945 managers gazed bleakly on the half-empty store and unfinished foundations. As postwar Seattle expanded in its direction, severe price-cutting disposed of the last lots. Alderwood Manor became a Seattle suburb.

from 7.3 billion to 2.2 billion board feet. Shipments from Saint Helens plunged from 113 million board feet in 1928 to 25 million in 1931. Half of Washington's sawmills failed. Western mill towns and logging camps vanished. By 1932 half the loggers and sawmill workers in Oregon and Washington were jobless.

On November 5, 1929, seventy-two-year-old William H. Talbot died. Rainier Investment Company president George A. Pope Sr. assumed the presidency of the timber firm. After forty-nine years a Pope was again in charge. Less forceful than his predecessor, the sixty-six-year-old Pope was nonetheless determined to secure a greater voice in McCormick Lumber. Previously, he had mainly dealt with investments, leaving Puget Mill Company affairs to Talbot.

Interested much more in ships than in lumber, Pope had joined the McCormick Shipping Company board in 1927. The 1920s had been kind to the maritime sector. McCormick's flag flew prominently in coastal ports and up the Columbia River to Portland. Two hundred and fifty of the vessels crossing the bar into the great river in 1928 were McCormick-owned. The firm, responsible for 20 percent of the Columbia's trade and ably managed in the Northwest by Hillman Lueddemann, topped all carriers. At first the coastwise business benefited from the Depression, as many shippers abandoned the railroads to secure lower freight charges on the water. The brief boom, unfortunately, attracted additional shipping to the coast, eventually weakening rates. By 1932 rates were a third of those charged in 1910. The next year coastwise and intercoastal freight tonnages dropped drastically. The McCormick Shipping line lost money in all but two years between 1932 and 1939. The Pacific Argentine Brazil subsidiary was also deeply troubled. Only bicoastal maritime subsidiaries, including the West Coast's biggest longshore operations, showed some profit. And low vessel prices made it worthwhile to add five modern bottoms to the thirty-ship company fleet.

REGAINING THE COMPANY

ON DECEMBER 8, 1931, THE FORMER OWNERS OUSTED CHARLES R. MCCORMICK FROM THE boards of both the lumber and the shipping concerns. Reluctantly carried out, the decision amounted to a bitter moment for the corporate founder. Hauptman's tenure in the presidency, meanwhile, lasted only until 1933. As a practical matter, George A. Pope took control of the steamship and lumbering firms. McCormick might have saved himself, except for the mounting debt and the impact first of the lumber recession and then of the Great Depression itself. The Popes, the Talbots, and the Walkers had loaned additional money and avoided foreclosure for as long as possible. Afterward, they actually possessed a firm larger than the one sold in 1925, though much diminished in its principal asset (timber), plus shipping and related facilities.

Unless finances were restored and costs reduced, the Popes and the Talbots believed, outsiders would in turn take over McCormick Lumber. They must therefore find means of paying off a pressing portion, specifically $2,472,500, of the first mortgage bonds. They intended to reduce outstanding notes on a year-by-year basis after that. In 1932 shipping executive Lueddemann was ordered to cut mill and other expenses to the bone.

Lueddemann reduced the enormous mill inventories by slashing prices, earning $400,000 in the process by fire sale tactics. Under his direction McCormick Lumber entered a sales agreement with three other western mills. The Pacific Atlantic Lumber Company (PALCO), formed under this arrangement, jointly retailed products in the East. In time PALCO controlled a quarter of the trade between the two coasts.

McCormick workers use a shrieking, unprotected cutoff saw to make mining timbers at a company dock in the late 1920s.

Lueddemann kept the Saint Helens mill closed for seven months. In San Francisco he shuttered the old Pope & Talbot lumberyard and leased the site. Camp Talbot was shut down and all the logging equipment sold. Somehow the business office squeezed a million dollars out of overdue accounts in 1934.

In 1935 management insisted that Alderwood Manor residents resume previously suspended payments on their farm sites. More than a hundred properties reverted to the company when their occupants moved out almost overnight. The tracts were rehabilitated and resold. At the upscale Broadmoor development in Seattle, in contrast, the firm was reluctant to institute foreclosure. Many residents were allowed to retain their homes until the economy improved.

McCormick Lumber retained twenty-five hundred employees in 1934. Avoiding layoffs as much as possible, management canceled shifts, shortened hours, and spread out work. Production was lowered and major expenditure avoided. Port Ludlow deteriorated. Rotten flooring in the mill caused accidents, and the uneven level threw machinery out of alignment. The antiquated steam mill all but sank into the bay. When the employees considered a strike, the company closed Ludlow in December 1935, putting 450 people out of work. Mill equipment and scrap were quickly sold, realizing $100,000. The closure stunned Jefferson County, as half of the town's population departed. Usable structures and gear were transported to Port Gamble by barge. The remaining dwellings were torn down for firewood and other salvage purposes.

The firm mandated deeper cost-cutting in the spring of 1936. Lueddemann and Charles L. Wheeler, executive vice president of the steamship and lumber companies, implemented the policy, closing the remaining logging camps and assigning the work to contractors. Selling two hundred miles of rails, ten locomotives, one

TOWING ON PUGET SOUND

"THE PIONEER DEAN OF PUGET SOUND TUGBOATS, was making Quilcene bay in Hood canal for a midnight tow. Out of the murk sounded a clanking of chains and chesty shouts. Black water underfoot. Black weather above. . . .

The captain [Teddy Charlesworth] had big logs tonight in a 10-section raft. The mate and his men, for all the darkness, trotted over the sticks like ponies over turf. . . .

This raft was in 10 square sections, with some 50,000 feet bound in each one. . . . This was an emergency run. The mill was short of logs, so the Pioneer was beating down with only 10 sections, when the rule is to make up four times as many from the booms. . . .

This is the story of towing logs on Puget Sound—a battle with wind and tide. . . . Everything is left to the judgment of the towboat skipper. He must balance the risk of piling up a raft on the beach against the need for logs at the mill he serves. . . .

The toughest tow on the sound is through Deception Pass, where a 12-mile per hour tide roars between Fidalgo and Whidbey islands. These northbound tows . . . wait for the exact combination of wind and tide which will let them make the mile run through the pass. Tows have waited there for three weeks. . . .

Many a tow, however, has piled up on these rocky shores. As many as 10 boats are sometimes lined up in the bay with their rafts, waiting for a favorable chance to charge through the pass. . . . Whole rafts have piled up on Strawberry Island with a thunderous crash of big butts splitting into slivers on the sharp rocks. The salvaging of such a wreck may take weeks, and the log pirates gather up the strays. . . .

[Chief engineer George Primrose] has seen the old tug through some close calls in deep water. Once in a furious Alaskan storm a great sea broke over her, and she was eight minutes righting herself and fighting to the top. On a double-tow job down the coast in wartime she fought heavy weather. . . . One moment the lead boat would seem to be far under the Pioneer's bows, in the next she would be high up, her propeller churning the spume.

'Every winter trip it was like we were on a giant teeter-board,' said the chief, 'with the Pioneer on one end and the lead boat on the other.' . . . But even in the comparatively smooth waters of Rosario and Juan de Fuca straits the swells often dip sections of log rafts out of sight from the towboat decks. . . . When a wind kicks them into whitecaps the inert logs seem to come to life. They bob and roll, they bump and grind. Heaving against the swifters, the butts squirt jets of water as high as five feet above the raft."

—from James Stevens, "Towing on Puget Sound,"
Four L Lumber News Letter [1920s] from a manuscript
in the Port Gamble Historic Museum

The Puget Mill Company and other mills jointly operated their first paddlewheeler, the *Resolute*, in 1857. She blew up and went down with four of her six crew in 1868. Port Gamble's best firs went into the company's *Cyrus Walker* in 1864, and its beams and planking into its first coal-burner, the *Tyee*, in 1884.

The twenty-one-year-old *Goliah*, the first ship of that name, entered company service in 1871, captained by the trumpet-voiced S. D. Libby. She was the company's second tug built in the United States. Libby was a renowned competitor. Hovering off a vessel, he outshouted and underbid the wiliest captains. (Competition ended when mill owners in 1891 formed a towing monopoly on Puget Sound.) William Gove ascended from mate in 1865 to captaincies of three successive company tugs. His Seattle obituary on March 6, 1912, honored his forty-seven-year career without "serious accident, although many of his acquaintances in the service met with disaster and mishap." Like many tug masters, Gove came from Maine.

For years mill expansions added tugs. The most distinctive was the *Polly*, a grim-looking former Russian gunboat slung low between great paddlewheels. The firm ultimately operated a big tugboat fleet. The 166-ton *Pioneer* was acquired in 1892. Four-foot-long seaweed covered her decks when James Stevens depicted her. The 65-foot, log-towing *Pioneer* replaced her in 1951.

Steam tugboats were practically indispensable for moving sailing vessels, upwards to three at a time, on Puget Sound. Dependent on them for transport, early settlers usually regarded the hardy tugs with affection. For the mills tugs saved on time and sails, carried executives and timber-finders around the Sound, and reduced groundings and wreckage. They also allowed cheaper recruitment and releasing of crews in Port Townsend: Towed vessels required only skeleton crews.

If able, ship captains resisted the expense of tugs. Off the California coast, Pope & Talbot's 1,224-ton lumber carrier *Enoch Talbot* drifted dangerously toward shore while her master haggled over towing price. He acceded to a rumored thousand-dollar towing fee once he "became convinced that it was a case of tow or wreck," reported the *Weekly Times-Telegraph* on July 25, 1895.

Tug skippers, ship captains, and loggers were proud, independent breeds generally disdainful of one another. The company tried to bridle independent skippers. Edwin G. Ames let recalcitrant vessels refusing tows to lie becalmed for days in Puget Sound. Their frustrated captains had laboriously to carry on under their own sails.

Towering deck piles of mill slabs fueled company tugs for decades. In 1885, Cyrus Walker worried that he would never find substitutes for wooding at any hour night or day if the Knights of Labor forced replacement of the Chinese dock employees. (The union failed.) Then, tugs and ships grew larger and more powerful, and coal-fired ones replaced the old vessels.

The Puget Mill Company always viewed tugs as supplementing mill purposes. In 1891 it established the Puget Sound Towage Company exclusively to tow logs. In a noncompeting pool arrangement that year, the firm and three other tug-owning Puget mills also formed the Puget Sound Tug Boat Company and added more tugs. A wholly owned subsidiary after 1905, it ended cutthroat competition and rate-cutting and provided better service. It controlled smaller tugs through the Admiralty Tug Boat Company after 1902.

Around the turn of the century, steamships greatly limited the need for the old sidewheelers and sternwheelers. The company's coal-fired tugs largely turned to the slow towing of heavy lumber barges to Hawai'i and between Puget Sound and the Columbia River, Alaska, or California. And they towed unwieldy log rafts from the Sound to California. Because windjammers remained in the world lumber fleet into the 1940s, Puget tugs also handled their greatly diminished number.

A noncompetitive pool formed in 1891 by the Puget Mill Company and nearby mills dictated Puget Sound tug prices to ship captains. It operated the *Wanderer* and other tugs, and in 1905 made the pool a company subsidiary. (Jefferson County Historical Society, Port Townsend, Washington)

The Medical Arts Building, seen in the 1930s, was one of the valuable buildings that Cyrus Walker, Edwin G. Ames, and other Puget Mill Company executives constructed during decades of land accumulation in Seattle. *(Todd Hazen, photographer)*

hundred railcars, and extensive logging equipment, the company secured $350,000. On the Atlantic coast, where McCormick Lumber was the sole remaining participant in PALCO, swollen inventories were liquidated in Albany, Boston, Brooklyn, and Philadelphia. Marketing charges were reduced by negotiating exclusive sales contracts with wholesalers, a few large industries, and the railroads.

The Depression and the subsequent world war revealed that American corporations could no longer do business in the old way. The worldwide economic collapse closed historical export markets. The outbreak of war in Asia in 1937, followed by the Munich Crisis of 1938, further dampened overseas trade. Pope and his associates, moreover, could no longer pretend to be heroes of capitalism. The climate of public opinion had changed, and corporate leaders were now widely considered to be suspect characters. Republican Popes, Talbots, and Walkers worried that the federal government had become, under the New Deal, too big and intrusive. Government support for the unionization of stevedores and sailors deeply angered the executives most closely involved with shipping. In the mills and in the woods the company confronted effective union organization efforts for the first time.

In the mid-thirties industrial and interunion warfare broke out in the Northwest camps and mills, spreading to nearby towns and cities. Owners clashed with unions, strikers with the police, and rival American Federation of Labor (AFL) and Congress of Industrial Organizations (CIO) organizations with one another. In May 1935, McCormick Lumber made a precedent-setting bargain, recognizing the conservative AFL-affiliated Lumber and Sawmill Workers Union and agreeing to the forty-hour week and a 50 cents an hour base wage. (Previously, laborers received $9 a week, with $6.75 deducted for room and board.) The mills subsequently developed good relations with the union.

Strikes organized by the CIO nonetheless closed almost all Pacific Northwest mills in 1937. As winter approached, the Port Gamble local appealed for a temporary reopening, so that people might at least earn enough money to eat. After some deliberation, management restarted a single shift. The company and the CIO sawmill workers thereby established a significant measure of mutual respect.

Labor troubles dogged Depression-era shipping interests. McCormick Shipping executives fiercely opposed the seamen and longshore unions' efforts to unionize, raise historically low wages, and challenge managerial prerogatives. Their 1937 annual report claimed that prolonged strikes explain "to a large extent the difficulties of doing business during 1934, 1935, and 1936." Worldwide shipping then slumped further, tying up important company vessels. More strikes loomed.

Despite continuing labor troubles, McCormick Shipping maintained a strong enough credit rating in 1937 to borrow $900,000. With the loan the concern bought seven McCormick Lumber vessels. The transaction enabled the lumbering firm to redeem remaining first mortgage bonds. The threat of outsiders taking over all but vanished. By the end of the decade, McCormick Lumber loans were paid off and the burden of remaining mortgage bonds was largely reduced. Between 1925 and 1937 corporate assets had plummeted from $25.4 million to $9.2 million. Port Ludlow, the logging camps and equipment, the lumberyards, and most of the timber were gone. In 1938 the Puget Mill and Rainier Investment Companies brought a final foreclosure action against Charles R. McCormick. On May 23 the court accepted their $6.1 million bid. Unable to pay obligations in excess of $8 million, McCormick was forced to cede all his holdings, including the McCormick Steamship Company.

In California the fully restored traditional owners incorporated the Pope & Talbot Lumber Company on June 4, 1938, to handle the firm's nonmaritime properties. Anticipating orders from Europe, Pope & Talbot added equipment at Saint Helens, raising production there from 84 million feet in 1938 to 111 million board feet in 1939. McCormick Shipping, kept intact for the time being, labored hard to secure a new federal subsidy. The government, unfortunately, rejected the request and for good measure took additional actions harmful to the firm's South American business. The directors promptly determined on partial liquidation of McCormick Shipping, selling ten vessels during 1940 and putting the remaining sixteen out to charter.

On June 29, 1940, Pope & Talbot, Inc., was incorporated and both the McCormick Shipping and the Puget Mill Companies formally dissolved. A new Pope & Talbot division under the old name, the Puget Mill

Company, handled real estate. Pope & Talbot Lumber managed the mills and the timber. A third division, retaining the McCormick designation, took on responsibility for the ships. George A. Pope Jr., now thirty-nine, assumed the presidency; his seventy-six-year-old father became board chairperson. Cyrus Walker's son Talbot, the president's brother Kenneth, and William H. Talbot's son Frederick accepted vice presidencies. Collectively, they restored the family tradition of preparing sons for corporate leadership. The junior George Pope had started as a helper in the sawmills, the others as assistants in a number of the concerns. All had now climbed to responsible executive positions.

Before the incorporation George A. Pope Jr. and Frederick C. Talbot met privately in a San Francisco restaurant. Pope made a forthright declaration: "I've got more votes than you have." As the result of his father's 1930 marriage to Harriet Brownell, the granddaughter of William H. Talbot's sister, Pope controlled the important block of stock belonging to Brownell's mother. He also held proxies from a sister and from some of the Talbots. "I'm going to run this company," he essentially told Frederick (Fred) Talbot. "I'll always treat you well. You'll always have a job. And you'll always have a say. But you have to shake my hand now, and say you recognize me as the leader of this company." Talbot accepted the arrangement, thereafter serving as senior vice president and remaining a faithful partner. As the United States moved closer to war, a new generation took charge on the Pacific coast.

ADVENTURE BY STEAM SCHOONER

"GONE ARE THE DAYS WHEN LUMBER SCHOONERS *made their way northward [from California] to outside ports either entirely empty or, at best, with just a few tons of supplies for the lumber mill. Today big business efficiency has stepped in and the lumber schooner goes north laden to the Plimsoll mark with general cargo and comes south [from Washington and Oregon] with general cargo, entirely surrounded by deck loads of lumber....*

On the forward deck [of the McCormick's Point San Pedro*] were boilers for Alaska canneries, parts of huge cogwheels for northern mines, heavy pipe-like cores for rolls of fibre board. The after deck was covered with 100 tons of huge bales of scrap paper, headed back to the paper mills to be transformed into cardboard ... Below, the hatches were filled with general cargo.... Two thousand cases of canned goods, 4000 packages of dried fruits and 400 sacks of sugar.... [Fourteen] tons of whisky, 650 cases of gin and 900 cases of wine explain why they also carried seven tons of coffee (black) and countless packages of remedies.... Mixed among automobiles, rice, sacks of lime, shovels, mining machinery, glue thinners, insecticides, ground bones and washing powders were items that spoke of strange ports in other seas....*

Southbound, the lumber schooners try to live up to their names. Don't be deceived when you see them steaming into San Francisco with their 10- to 14-foot deckloads of lumber, however. That's just camouflage. Oh, they might have a million and a half or even two million feet of lumber on decks and below, but they will also carry matches, caustic soda, paper pulp, box board, structural steel, scrap tin and carloads of flour."

—from Theodore Smith, "Adventure by Steam Schooner," *San Francisco News,* March 23, 1937

Workers load lumber onto the Point San Pedro at the Saint Helens mill dock.

CHAPTER FIVE
REVITALIZATION

1941-1961

WARTIME

POPE & TALBOT, INC., BEGAN WORK IN A TIME OF UNCERTAINTY. GEORGE A. POPE SR., chairperson of the board, named his son president in 1940 but retained the real authority while waiting for his polo-playing namesake to prove himself capable of leading. Meanwhile, operations "were more or less of a liquidating nature," a later annual report recalled. "Vessels were being sold with no plans made for their replacement and our timber was being cut at a rate which would have depleted the resources behind our mills."

Managers ordered both mills to concentrate on production of cheap green (undried) lumber. At Port Gamble a new headrig was installed to reduce capacity and stretch timber supplies. Neglected for several years, the Saint Helens mill installed a new 10-foot band headrig and gang saws to increase capacity. A second sawmill, housing a double-cut band saw, started operations there soon after the attack on Pearl Harbor.

Faced with uncertain prospects, Pope offered the firm for sale, and a rich suitor came courting. The Aluminum Company of America (ALCOA) sensed a bargain and offered $8 million in 1941—about what McCormick once owed the restored owners. But after a tour of corporate properties, Hillman Lueddemann convinced the younger Pope that ALCOA's offer was far too low. Nevertheless, when ALCOA withdrew its bid, spirits temporarily plunged. The failed sale, in retrospect, happily maintained corporate independence, but the firm's immediate future seemed in doubt.

The situation became more grave when the board chairperson died in 1942 and George A. Pope Jr. entered the army. Talbot Walker, trusted but inexperienced, abandoned a leisured California existence and assumed the helm. He performed well until ill health forced his resignation in 1943. Executive vice president Charles L. Wheeler, an articulate promoter and a key figure in developing and reorganizing the McCormick Steamship Company and the Pacific Argentine Brazil Line, replaced Walker as chief executive.

The war brought enormous unexpected changes. A long-running economic boom came to the western United States. The last, and in many ways the greatest, of the westward migrations brought waves of newcomers to the booming industrial centers on the Pacific coast. New capital and new industry responded patriotically and profitably to the opportunities of wartime. Defense contracts stimulated aircraft and ship building, doubling and then tripling employment in regional manufacturing.

Wartime brought women into Northwestern sawmill production. Here, one woman and several men work on the "greenchain," probably at Saint Helens. They pull lumber from the conveyor and stack it by size, length, species, and other criteria.

Torpedoed to starboard, the *Absoroka* lost one seafarer on December 24, 1941, and limped into a California harbor. Once repaired, this intercoastal vessel was one of five company vessels remaining at the end of the war.

> "In 1940 my grandfather, George A. Pope, made one last try to sell Pope & Talbot, this time to the Aluminum Company of America (ALCOA). But Hillman Lueddemann convinced my father that Pope & Talbot's Northwest properties were worth much more than ALCOA was willing to pay. My father also managed to convince my grandfather that the ALCOA sale would be a bad deal. It never took place. I think this was the moment in my father's life when he determined that even though he had many other interests, he would keep Pope & Talbot so that the next generation would have the opportunity to work there."
>
> —Peter T. Pope

Every "major war development," the capable Wheeler observed in 1944, "is of striking and important value to the lumber industry." The war kicked off a lumber trade boom that lasted, with only occasional interruption, until the 1970s. Timber values escalated dramatically and ended the long era of relatively cheap trees and lumber. Sawmills and loggers recovered from the Depression. Throughout the West, opportunities brightened for impoverished lumber towns.

Faced with war needs, the federal government took effective control of raw materials, production, and prices. This emergency wartime policy, not just company interest, motivated Pope & Talbot. The company eagerly supported the national goals of all-out production and restrictions on civilian business. Federal manpower officials helped the firm to stabilize its workforce. In response to governmental expectations, the Saint Helens mill shifted from ship to rail transport. Unrestrained private and public timber cutting kept resource-pressed mills running. The value of standing trees doubled and even tripled. Company forests increased substantially in value.

With demand suddenly so great, Pope & Talbot struggled to find workers and exploitable timber. Employees who entered the armed forces had to be replaced. And the labor shortage was compounded by all-out production. Twenty-four-hour-a-day operations could be maintained only by constant hiring. Women secured blue-collar jobs for the first time in the Saint Helens mill. Transients were placed on the payroll. Logging crews were hired—their numbers expanded by the need to cut timber as fast as possible to feed the hungry mills. The firm built all-weather roads and transported loggers between the woods and towns. Trucks and diesel-powered crawler tractors capable of yarding the biggest logs in the roughest terrain were acquired, increasing efficiency and raising employee productivity.

Largely because of Saint Helens, Pope & Talbot retained its status as a leading Pacific Northwest lumber producer. The Saint Helens operation outproduced all other Oregon mills during the Second World War. Annual output reached 194 million board feet before declining to 109 million in 1946, the first full year of peace. The mill was awarded a coveted Army-Navy Production Award in recognition of its contribution to the defense effort.

The government also figured largely in the growth of Pope & Talbot's shipping business. Federal officials depended on the concern's expertise in maritime matters. Company terminals handled millions of tons of war cargo each week. The War Shipping Administration assigned the firm operating responsibility for

PUGET SOUND SAWMILL SYMPHONY

"THE HIGH SCREECH OF THE BAND SAWS IN A hundred pacific Northwest mills is a battle song. The slam and thud of logs on a sawmill's carriage, the groaning of the bull chain, the rattle of lumber on metal rollers, are undertones in the raw-melody singing of pontoon planks across which tanks will rumble, and of dock timbers for invasion beaches scattered around the world....

By truck, rail, car-barge and ship, the [Port Gamble] mill's capacity output of approximately 175,000 board feet a day is pouring to defense points in this country and to the fighting fronts....

General superintendent of the mill is chunky, lumber-wise W. N. Hammers[ch]mith, [who said,] 'Everything's destroyed when the Army goes into a place now. The lumber ships go right in behind the assault troops so the engineers can start rebuilding.'...

The bull chain hurls the logs up the slips [from the log pond] and into the thundering cavern of the mill itself. There aren't any spindling noises in a sawmill. They're all heavyweights. The log booms down onto the carriage with a rolling crash. Three doggers [bite in to] secure it. The sawyer measures it with his eye—he can tell you to the odd foot how many sticks he's going to get out of it and what size they'll be.

The carriage rambles into the saw with the rush of an express train. The saw hits the log, lays bare a gleaming slice of it with a stormy whine. That double-edged band saw is 61 feet long, travels 10,000 feet a minute. The carriage roars back, and another slice comes off. The slabs go bouncing and clattering along rollers, on their way to the trimmers.

Sawyer Walter Hirschi stands eagle-eyed at the levers operating the carriage, engulfed in a torrent of sound. He's shielded by a screen from the sawdust flung up by the spitting saw, but some of it filters through to hang on his hat, his shoulders, his eyebrows. 'I used to think,' he remarks, 'the sawyer had the best job in the mill, but he does just as blasted much work as anybody else around here.'

As vital as the sawyer is the saw filer. He's W. A. Thompson, and he's been filing at the Pope & Talbot mill for 51 of his 70 years. He's proud of the fact that his father, an old Maine filer, came to Port Gamble for Pope & Talbot in 1858. 'The teeth have to be absolutely sure or your saw will lead one way or the other and you'll get wavy lumber,' he explains. 'And the points must be kept wider than the blade. If the blade gets to rubbing the log it heats up and is apt to crack and explode.' There's more to it that Thompson can't tell. When a man has a 'feeling for saws,' he's a filer. If he hasn't got it, nothing can ever make him one.'"

—from Robert Mahaffay,
"Puget Sound Sawmill Symphony
Heard from Naples to Rabaul,"
Seattle Times, February 27, 1944

This account, with the patriotic tone common to much of the wartime media, stressed the mill's contributions to far-flung combat zones. Actually, about 60 percent of Port Gamble's lumber, 125 to 150 million board feet, traveled short distances. It built naval housing complexes in Bremerton, in Keyport, and at other Washington bases. The reporter described the gang mill installed in 1925, when the new plant received improved saws and lumber-handling devices, kiln dryers, shields, doggers, and other safety equipment. He focused on the initial steps, the massiveness and sounds and on two skilled workers. He ignored the dampness and dangers. Saw-filer Thompson had pride of skill and strong family ties to the place. Hirschi, promoted to sawyer since his hiring in 1932, disliked the pace. He later complained to a grandson about supervisors always pushing him to speed up.

Skilled maintenance employees and processing workers were traditionally the best-paid in western sawmills. Filers were the highest paid. Most employees still filled unskilled and semiskilled jobs, however. Historically, they earned less and quit faster than skilled employees. Wartime pushed their base wage up sharply in the Douglas fir region. Mills paid laborers 50 cents an hour in 1936, 75 cents early in 1941, and 90 cents in 1942. In heavily unionized western sawmills, wartime employees earned even more on overtime beyond the eight-hour day and the six-day week.

A naval housing project in Bremerton, built entirely from Port Gamble's lumber, 1942. Some 60 percent of the mill's wartime lumber went for housing on several Washington naval bases. Civilians got about 10 percent of all of the mills' wartime output; the armed services got 90 percent.

seventy-six vessels. These ships called at such newly famous destinations as Bizerte, Guadalcanal, Murmansk, and Salerno. Four ships went down, including the *West Ivis* with all hands. Another was incorporated into the artificial harbor assembled to support the invasion of Normandy in June 1944.

Big-government insurance policies on ships, cost-plus contracts, and generous tax write-offs guaranteed profit to wartime business and industry. From 1942 on Pope & Talbot after-tax profits averaged $2 million a year. The "war sort of saved the company," Peter T. Pope has said, "as far as building up cash and giving it a future."

PROSPERITY

BY THE END OF THE WAR IN 1945, POPE & TALBOT FOUND ITSELF AT ONE OF THE greatest turning points in its history. A company revitalization and reorganization was needed to address the needs of the future and to manage the transition from production for war to production for peace. The war had pulled the country out of the Depression and the company out of the doldrums. With company morale restored, selling the firm was now unthinkable. An invigorated George A. Pope Jr. returned from the army ready for fresh initiatives and keen to be an active leader and corporate builder. Marine veteran and Harvard M.B.A. Adolphus Andrews Jr., who was married to Pope's niece Emily, joined the firm in 1949 and was elevated to the board in 1957. Pope wanted him to gain a major voice in corporate affairs.

Fresh long-term plans and growth-oriented spending marked the immediate postwar period throughout corporate America. Pope & Talbot's leaders devised a ten-year, $26 million expansion and diversification strategy. Pope relied heavily on nonfamily executives. A new vice president, the experienced shipping executive E. N. W. (Ed) Hunter, took control of postwar maritime activities. Wheeler and former captain Hunter, an Annapolis graduate who liked to run a tight ship but initially knew little about the lumber industry, advocated and guided the steamship buildup. Lueddemann turned the company toward forest and mill growth.

The flourishing American economy contributed in a big way to corporate expectations. The United States entered a long and steady wave of growth and prosperity, and the powerful economic tide found Pope & Talbot well positioned to ascend with it. For two decades America's output and demand increased in powerful tandem. Median family income rose by 40 percent in the 1950s and again in the 1960s. The rate of inflation, meanwhile, remained relatively low. High consumer demand, cheap energy, government-financed technological advances, and military and foreign aid expenditures had a significant impact on Pope & Talbot.

Flush with wartime earnings, company leaders responded aggressively to postwar opportunities. As the value of raw materials soared, the pressure to use more of their timber—and add forests—escalated. In a short time the firm acquired a vast tract in Oregon and opened tree farms. It diversified into new products and markets as opportunities presented themselves. Unproductive operations were shed. Administrative reforms were implemented in a quest to more efficiently use resources. Although Pope & Talbot was not a party to the great lumber mergers and acquisitions of the 1940s and 1950s, it was affected competitively by the trend, creating a spirit of grow or die. From such amalgamations Georgia-Pacific and United States Plywood joined Weyerhaeuser and Pope & Talbot as the principal Douglas fir producers.

Company leaders recognized the vast opportunities generated by postwar changes in America. A booming population coupled with robust construction and defense industries pushed demand for timber

A lunch at the Fairmont Hotel in San Francisco celebrates the company's centennial. Pictured clockwise from front left: Emily Pope Montgomery, [unknown], Violet Dutton, [unknown], Ed Hunter, Bill Brownell, Dolph Andrews, Winnie Wilson, Jeff Montgomery, Sophia Hutton, Cy Walker, [unknown], Vera Michelson, Vera Talbot, [unknown], George Pope, Helen Gibbs, Talbot Walker, Sophia Talbot Brownell, Emily Andrews, Ted Talbot, Katy Brownell, Charlie Wilson, [unknown], Dr. Lewis Michelson, Amylita Talbot Wilson, Alex Baldwin, [unknown], Mary Hunter, Charlie Wheeler, Zib Pope, Fred Talbot.

At the hundredth anniversary of Port Gamble in 1953, the individuals who led Pope & Talbot during and since the war pose for the camera: Talbot C. Walker *(right)*, George A. Pope Jr. *(center)*, and Frederick C. Talbot *(left)*.

PROSPERITY **89**

AFTER THE WAR: WHAT NEXT?

"IT IS A PECULIAR FACT THAT EVERY MAJOR WAR *development is of striking and important value to the lumber industry, appraising it from a post-war viewpoint. In that respect, let's examine what the fortunes of this war have done for our lumber industry here in the Pacific Northwest. In the face of wartime needs, and consumption exceeding production, there is simultaneously accumulating a vast potential need for lumber and lumber products for the construction of rural and urban dwellings, farm buildings, school buildings, highways, railroads, rehabilitation, and for numerous other types of construction that have been deferred until the war is over. In the embattled areas the destruction is bound to absorb great quantities of lumber for years to come. . . .*

Most remarkable is the coincidence that war has been provident to our industry. . . . It has caused the construction of thousands of ships of a type that will transport lumber perhaps better than any other commodity to the markets of the world.

The war has rushed to completion more power plants so that we shall have a plentiful supply of low-cost power. It has brought hundreds of thousands of new workers to the Pacific Northwest. . . . [In peacetime] we should have low-cost equipment, a reasonable supply of competent labor, and a stupendous demand for our products. . . .

Our industry is indeed fortunate in not having to face the serious problems of reconversion that confront so many other industries. We have only to re-enter familiar markets, and exert ourselves a little to seek new outlets as well to expand old ones. . . .

There are a number of steps the industry could take that would [adhere to the proverb] in time of plenty prepare for times of stress.

1. The industry should more aggressively strive for maximum utilization through intensified development of plastics, pulp, veneers, lamination, prefabrication, distillation, preservation— and many other methods—of the timber now remaining in the woods. We should seek the assistance of our Pacific Northwest universities and similar institutions, and we should obtain the services of the BEST *industrial, chemical and forest laboratory engineers available.*

*2. Starting immediately [we should] expand our cooperation with county, state, and national authorities to achieve sustained yield of that great crop—*TIMBER.

3. We should withhold criticism of others until . . . we have eliminated the archaic methods within our own industry— for instance, let the world look to a SINGLE *standard inspection bureau.*

4. Immediate plans are needed for vigorous support . . . to market our products through the world; and to simplify or revise market distribution domestically. . . .

5. Lumbermen individually should support constructive post-war controls to assure economically sound prices and volume of production.

6. A strong post-war economy is necessary to keep business healthy and the keynote thereof is plenty of employment. . . .

As an industry we must support post-war research. . . .

You must have a long-range vision to see that if lumbermen continue, at their present rate, to leave 42% of forest lands unproductive after harvesting the virgin forests, they are not building toward a healthy future. . . . The tree farms we've heard about are a splendid development. . . . [It must no longer be said,] 'Most trees wind up with one-third left in the forest and one-third left in the sawmill.' . . .

We . . . should cease our criticism of the consumer, labor and our Government until we have discontinued some of the archaic, bureaucratic and parasitic practices indulged in by the industry. . . .

Our industry needs an outlet for at least 20 per cent of production in foreign markets and . . . some mills might desire as high as 60 per cent of their production being sold offshore. . . .

We hear of increased pulpwood *drain, in the future, from the forests. The shift of economic conditions will probably bring about a great utilization of the hemlock, spruce, and fir in the Pacific Northwest. But Canada has about 70% of the spruce, fir and hemlock yielding sulphite and mechanical pulps. Since the United States is dependent on Canada for this readily available pulpwood, harmonious cooperation must be maintained....*

Not only are we morally obligated to find jobs for millions of these soldiers when they are demobilized, but also we must reabsorb millions of displaced war workers....

[It] is reasonable to assume that at war's end, we will have the merchant marine to handle any volume of export trade that develops.... Be forewarned about [foreign countries rebuilding fast and offering cheaper transportation], and stand as a group for 'Ship American,' no matter what comes.

The post-war merchant marine will be a huge fleet of modern freighters and the backbone of this fleet will be the 15-knot and up Victory ships. With this speedier vessel our lumber products will reach foreign markets through frequent and regular sailings.

Coastwise and intercoastal lines will receive their replacements quickly since these services were the first to be denied their ships when hostilities broke out. The shipping companies are modernizing their methods of handling cargo and keeping up with war-produced improvements at ports."

—from Charles L. Wheeler,
"Industrial Planning for the Post-War Period,"
speech [c. 1944], from a manuscript
in the Port Gamble Historic Museum

Late in 1944, acting company president Charles L. Wheeler gave a insightful trade association speech, from which selections appear above. He outlined much of what Pope & Talbot planned—and did—after the war. Wheeler was a well-known "shipping man" and an articulate promoter. He had been a key figure in developing and reorganizing the McCormick Steamship Company and its subsidiary, the Pacific Argentine Brazil Line. In 1935 he became executive vice president for steamship and lumber operations.

Wheeler proposed few sentiments or actions disagreeable to lumber firms with large timber holdings. That sector looked forward to a bright future of tree farms, low peacetime reconversion costs, and modern scientific and technical advancements. Its leaders disliked antiquated methods in the industry. They happily anticipated pent-up domestic demand and substantial world-market possibilities without significant foreign competition, at least for a while. They appreciated that the war and a beneficent federal government had brought prosperity to their depressed industry. Unlike many Americans, Wheeler anticipated postwar growth, not a depression, and planned accordingly.

Notably, he cautioned against criticizing consumers, unions, or federal policies. He even endorsed retaining price and production controls after the war—views popular with consumer advocates and New Dealers. These sentiments came from somebody who had fiercely battled prewar maritime unions and despaired of prewar New Deal regulations. Industry attitudes seemed to be changing.

Patriotism and self-interest encouraged Pope & Talbot to cooperate closely with Democratic-controlled administrations. "Government and industry have been getting closer and closer together," Wheeler wrote Frederick C. Talbot on June 23, 1949. "We believe that the government has gone a long way in the last ten years towards meeting some of the problems of the private lumber industry."

During that period national forests had been thrown open to loggers, and government cargoes had filled ships. Congress had given the lumber industry at least three legislative plums as well as a recent major law restricting unions. War-built vessels had been transferred at low cost to Pope & Talbot and others. Federal subsidies were keeping its South American line steaming. And consumer spending, partly fired by the Cold War, was paying dividends. Wheeler had good reasons for optimism.

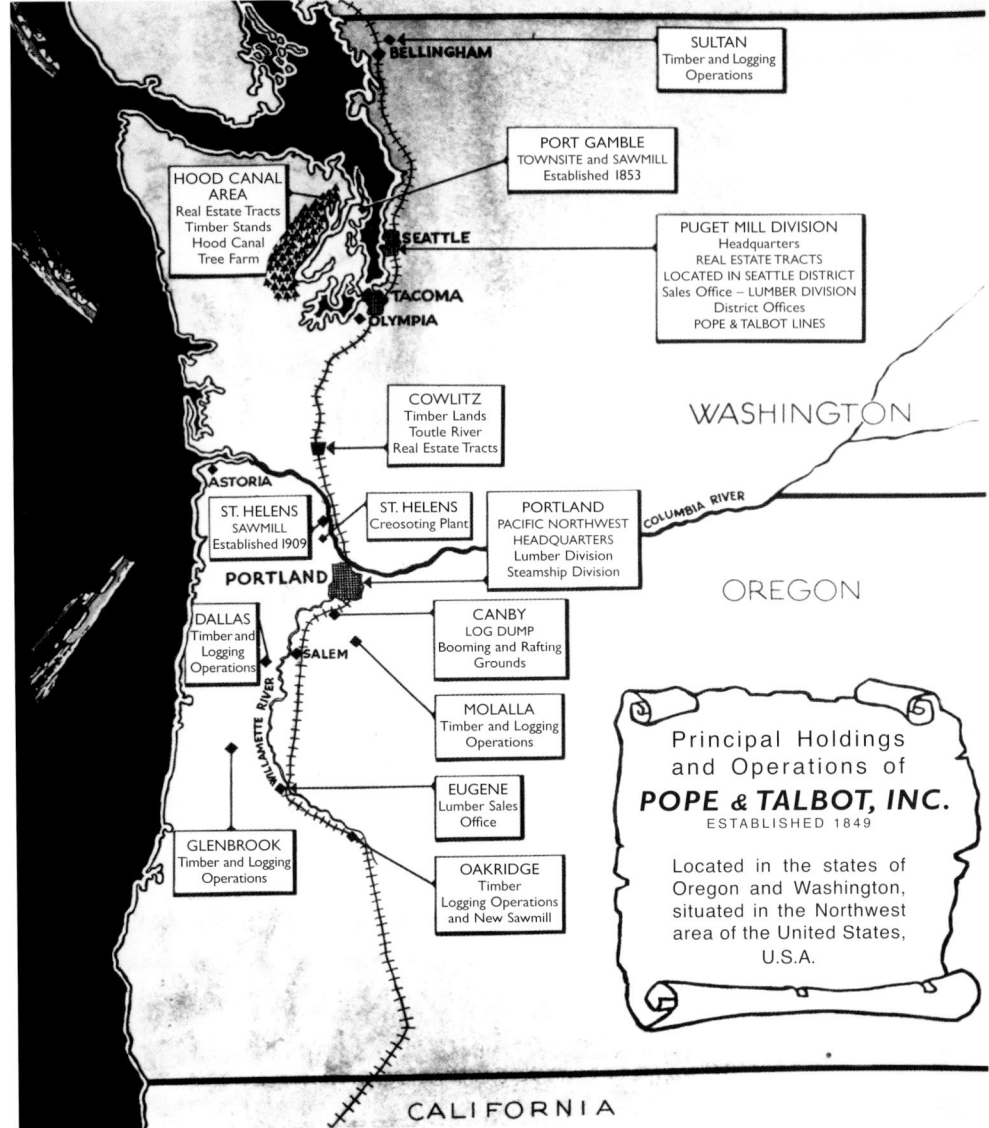

The company's principal postwar holdings.

products. Fifteen million housing units, 80 percent of them single-family dwellings, went up between 1945 and 1955 alone. The federally subsidized building boom had no end in sight. To Pope & Talbot, growth meant lumber sales, maritime traffic, and the transformation of cutover land into Washington subdivisions. The reconstruction of war-torn Europe provided profit-making opportunities overseas.

The company still operated in a transportation-sensitive industry, where shipping costs historically played an immense role in profit or loss statements. Although California was responsible for a fifth of the nation's population growth during the 1950s, it relied far more on nearby Oregon rail mills for material than on the more distant Port Gamble or Saint Helens operations. These two mills, to aid Pope & Talbot's maritime operations, filled company bottoms bound for the East Coast, where by 1956 a single New York yard handled most of their output. But before then the firm had built its own immense Oregon rail mill to reach the burgeoning California and Midwestern markets.

INTO OREGON

IN 1939, OREGON SUPPLANTED WASHINGTON AS THE NATION'S LEADING LUMBER-producing state. In his most important innovation, George A. Pope Jr. followed Hillman Lueddemann's postwar advice to follow the trees into Oregon. This turning point in the company's history required a series of major investments and new products. Pope initially arranged the purchase of the 32,000-acre Penn Tract for $2.5 million in June 1946. This savvy move was followed closely by the construction of a state-of-the-art mill in nearby Oakridge. Company forester Loran "Stub" Stewart had ignored industry scoffers, who dismissed the tract as too filled with "decadent" trees to be commercialworthy. Stewart estimated that a billion-

Two loggers on an early motorized crosscut saw cut one of the Penn Tract's trees for the Oakridge mill in the late 1940s.

A U-shaped steel frame on wheels facilitates a crawler tractor skidding logs to a landing from the Willamette National Forest in the late 1940s. *(Harold M. Brown, photographer)*

and-a-quarter board feet of good timber grew among an array of those rotten at the core. Moreover, and of strategic importance, Penn Tract sections were 98 percent virgin Douglas fir and were intermingled checkerboard style with 160,000 acres of the Willamette National Forest, containing another 4 billion board feet that would, sooner or later, be sold to the highest bidder. Pope & Talbot assured itself of a long-term Oregon timber supply with the Penn Tract purchases. The company lived off Oakridge timber for forty years.

Postwar demand nearly overwhelmed private western timber inventories, already low from decades of heavy cutting and liquidation. The national forests were generally less productive than privately owned tracts but contained more standing timber. Only a small portion had been harvested by the end of the war. But this was about to change. The annual cut for all national forests rose from under 4 billion board feet to more than 9 billion between 1950 and 1960. In the quarter century after the war, federal land supplied 20 percent of the nation's wood needs. The harvest rate in many western states exceeded half the lumber produced. Public timber kept Port Gamble and Saint Helens running. Had another very large fir tract ever come on the market, buyer competition would have been intense.

OAKRIDGE

COMMERCIAL LOGGING INVOLVED INDUSTRIAL-STYLE LOGGING, FOR WHICH UP-TO-DATE equipment required fewer crew members than before the war. Powerful shovels and bulldozers cleared Penn Tract yarding and building sites, created fire protection, and opened 150 miles of logging roads by 1953. (Logging railroads were ruled out because many trees rose on Cascade mountain slopes into snow-covered elevations.) Fellers, working in pairs, used gasoline-powered chain saws to drop a giant tree in five or six minutes. Caterpillar tractors, big cranes, donkey engines, spar trees, cables, and newer devices filled the busy sites. Trucks carted the logs ten to fifteen miles to Oakridge, which was about forty miles from Eugene.

Under the leadership of Cyrus T. Walker, Pope & Talbot established the Upper Willamette Tree Farm on all but 2,000 acres of the Penn Tract. The new entity represented 70 percent of the corporation's timber holdings, trees destined at the appropriate time for harvesting. The tree farm movement was another domestic phenomenon of the war years. A cooperative industry nursery on Puget Sound had 6 million seedlings growing by 1945. Dedicated in 1947, the company's first Washington venture into tree farming covered 53,500 acres in Jefferson, Kitsap, and Mason Counties. "Farm" acreage was expanded two years later to 72,804 acres via purchase in Pierce County. Another 8,000 acres were added by 1956.

Sustained-yield methods, adopted for the most part on an industry-wide basis, guided foresters: clearcut select units, burn the rotten logs and slash, and replant and nourish the site. The main forestry problem was no longer how to best fall and move a 10-ton tree, but how to profitably grow and handle ten 1-ton trees in its place. Finding use for, rather than destroying, the rotten logs and slash was an overall corporate and industry problem.

As of 1948, Pope & Talbot planned to take 50 million board feet a year from the Upper Willamette Tree Farm and national forest, supposedly their sustained-yield capacity when cooperatively "farmed." Replacement stands were to be grown like crops and cut on rotation. Management felt confident in avoiding overhasty harvesting because fixed costs were low and the company had no financial problems. Timber would be cut in 2048 on land first logged in 1948. Sufficient timber would thus be available to maintain the Oakridge mill for at least

The massive Oakridge mill, 1949. A few years later the company would add a particleboard plant, green veneer mill, pilot plant for experimental work, and hardboard plant at the big Oregon sawmill. *(Howard Studio, photographer)*

Huge band saws are held in an upright position in Oakridge's saw filing room in 1949.

a century. One annual report optimistically predicted that it could be operated "at a fixed level, forever, and never run out of timber." The sustained-yield dream shattered in the face of competitors who blocked the firm from gaining exclusive access to this federal timber. In addition, the company had not anticipated changes in national environmental policies and their impact on harvest practices and availability of timber.

The company assured its future as a major producer with the opening of the Oakridge mill, its first lumbering plant in eighty years. This massive million-dollar affair turned the 1947 village of 650 residents into a town of 2,500. It built roads, recruited businesses, enlarged water and sewer facilities, and acquired and often sold housing sites, homes, and apartments. Its big, mechanized electric- and steam-driven mill could efficiently produce 200,000 board feet of lumber for every eight-hour shift. Trucks and railroads hauled the air- and kiln-dried output mainly to California and the Midwest. The company anticipated the addition of related manufacturing enterprises at Oakridge.

Managers initially rejected building a sash and door plant, a plywood plant, or a pulp mill at Oakridge. Instead, in the 1950s the company opened a particleboard plant, a green veneer mill, and a pilot plant to turn waste into a fertilizer-soil conditioner. Particle panels, made by intensely heating and pressing chips or shavings with resin, were then popular in the construction and furniture industries. The green veneer plant (and another in Port Gamble) debarked, sawed, and on gigantic lathes peeled little-valued cull logs against razor-

Trucks and freight trains served Oakridge and carried its products to the Midwest and elsewhere.

sharp knives. Thin sheets of wood peeled off and were clipped, graded, and sold to plywood and laminated wood manufacturers. Chip-n-Saw headrigs, once installed, converted the "peeler" cores into studs for framing building walls and landscape timbers. Pope & Talbot was transforming itself into a fully integrated wood products company.

"We are now entering into the fields of wood chemistry and wood utilization," the company announced. All of its mills must follow the "full utilization" doctrine. The "era of cheap stumpage, cheap labor, and seemingly endless forests" had ended. More careful timber harvesting and more efficient production methods were imperative. Management demanded maximum recovery from every tree and log—and not simply more product in general, but higher-value new products in particular.

In response to this directive, Oakridge and Port Gamble at first merely installed mechanical log debarkers. As new wood products emerged from drawing boards and laboratories, Oakridge initiated modest research projects. This applied research stressed improved wood usage and product development. Nationally, industrial scientists perfected particleboard, plywood, pressed fuel, and numerous pulp and cellulose variations.

By 1957 the Oakridge laboratory confirmed the benefits of building a particleboard plant. Sawdust and shavings once burned for heat or power there could instead be turned into particleboard, an economical substitute for plywood and lumber in construction and furniture making. Oakridge also constructed a green

Oakridge's headrig in 1948 easily handled the giant old-growth coming from the Penn Tract and federal sections within the Willamette National Forest.

98   REVITALIZATION

Sampling the moisture content in lumber was one of many quality controls in the new Oakridge mill. *(Harold M. Brown, photographer)*

veneer mill to use the otherwise unmillable decadent logs, and a pilot plant to turn sawdust into fertilizer. Another important innovation came in 1960. Adolphus Andrews Jr. had another mill facility test the manufacture of medium density fiberboard (MDF), partly using waste otherwise sold to pulp mills. The company led the industry as the nation's first MDF manufacturer for the furniture industry.

Timber reserves provided Pope & Talbot with a substantial raw material cost advantage. The rapid escalation of stumpage and log prices between 1947 and 1954 benefited old-growth owners. Management aptly termed their Douglas fir holdings as "golden nuggets." The company books, in standard industry fashion, had always carried timberland at cost, typically lower than current market values. After the war such tracts realized profits denied competitors dependent on open-market logs or recently bought timber stands or burdened with high-cost or poorly financed mills.

For the firm neither the tree farms, the distant Penn Tract, nor the removal of the last old-growth timber near the onetime Utsalady mill offset a growing shrinkage in the Saint Helens and Port Gamble cutting circles. Management struggled to keep Saint Helens going and to give Gamble a new lease on life. Both plants kept running only through open-market log purchases. Something had to change. Deciding that the mills must henceforth use smaller second-growth logs, the company, in a significant move, rebuilt the plants to handle younger trees and the lowest grades. Before then "nobody knew how to handle second-growth Doug fir," future executive Guy Pope said in a recent interview. In 1955 and 1956 it replaced high-speed gang mills with rigs designed for small logs and species other than fir. Profits rose from the higher wood recovery rate and broader lumber products. Port Gamble became a bright spot on corporate books, giving the company, according to Guy Pope, the initiative later to expand into small log country in the Black Hills, British Columbia, Oregon, and Washington. "We capitalized on the success of Port Gamble," he said.

SHIPS

THE COMPANY EXPECTED A ROSY POSTWAR FUTURE FOR THE SHIPPING DIVISION, ITS subsidiaries, and related businesses. New Pacific coast industries required huge quantities of bulk material, as did the European and Asian recovery programs. The East Coast demanded large amounts of lumber and other western products, and the West Coast presented a growing market for European and East Coast products. Commercial connections with South America were also expected to grow in importance. A further stimulus for the shipping division arrived when the government announced it would return vessels it operated during the war and also sell, at cheap rates, the enormous merchant fleet built up under federal auspices.

Pope & Talbot had to charter ships through 1947 in order to handle the peacetime intercoastal connection. Bulk cargoes, lumber making up 85 percent of the total, were discharged in ports from Massachusetts to Virginia. Steel and consumer goods filled the holds on westbound voyages. Chartered vessels also carried grain and coal to Europe and Asia. The firm reinstituted coastal service in 1946. In 1947 it assigned recently acquired steamers—the P&T *Seafarer*, the P&T *Forester*, the P&T *Pathfinder* and the P&T *Trader*—to the South American run.

American shipping concerns had their choice of modern vessels, courtesy of the war surplus disposal program, at prices as low as one-third the original costs. Pope & Talbot obtained and refurbished seven former troop transports, the largest and costliest vessels in its history, for a total payment of $8.6 million. Four prewar ships were sold for $8.5 million, financing the overall improvement. By 1950 the company and the subsidiary Pacific Argentine Brazil (PAB) Line owned outright ten fast steamers.

Despite initial optimism, the American maritime industry was destined to decline. No U.S. coastal or intercoastal operation regained its prewar strength. Between 1945 and 1950 American-owned tonnage dropped from three-fifths to one-third of total world tonnage. Increasing freight rates and competition from trucks, railroads, and less-regulated foreign-flag vessels, along with continual volatility on the labor front, compelled firms to either abandon the oceans or to take on foreign flags of convenience.

Heavy truck and train competition and an increasing wage burden drove Pope & Talbot from the coastal trade in the late 1940s and early 1950s. Although subjected to highly cyclical trade patterns and price cutting by rivals, the company—thanks to a federal operating-differential subsidy in 1948—still served South American ports. But American shipping firms were, as demonstrated by the southern trade, increasingly dependent on governmental preferences, subsidies, and cargoes. The company's lone surviving intercoastal competitor withdrew from the trade in the 1950s.

A three-month dock strike cost Pope & Talbot $1.2 million in 1948. Retrenchment followed. The firm contracted out stevedoring services, renegotiated terminal leases, dissolved two subsidiaries, and used

POPE & TALBOT LINES ROUTES

ROUTES

PACIFIC ARGENTINE BRAZIL LINE

Pacific Coast Ports to East Coast of South America. From Vancouver B.C., Seattle, Tacoma, Portland, San Francisco, Los Angeles, via Canal Zone to Trinidad, Rio de Janeiro, Santos, Montevideo and Buenos Aires...return same route, to Pacific Coast Ports...via Canal Zone. Dash-line on map shows alternate route.

PACIFIC & ATLANTIC INTERCOASTAL

From Baltimore, Philadelphia, Norfolk, via Panama Canal to Los Angeles, San Francisco, Oakland-Alameda, Stockton, Portland, Seattle and Tacoma...from Tacoma, Seattle, Portland, San Francisco, Oakland-Alameda, Stockton, Los Angeles to East Coast ports, via Panama Canal.

PACIFIC WEST INDIES · PUERTO RICO

From Tacoma, Seattle, Portland, San Francisco, Oakland-Alameda, Los Angeles to San Juan, Ponce and Mayaguez, Puerto Rico via Canal Zone.

Routes of postwar Pope & Talbot lines.

A Pope & Talbot terminal in San Juan, Puerto Rico, 1949.

chartered vessels to carry lumber eastward. The outbreak of the Korean War in 1950 stimulated shipping and lumber sales. During the hostilities Pope & Talbot operated a fleet of nearly thirty steamers on behalf of the government. But strikes afterward badly affected its steamship business.

Foreign-flag competitors finally ruined the South American connection. The company sold its four vessels in 1956 and distributed the proceeds to stockholders. George A. Pope Jr., typically saluted by newspapers as a shipping (not a mill) magnate, lost interest in Pope & Talbot as the historic shipping operation shrank. He did, however, remain interested in maintaining it uneventfully into the next generation as a family-led firm.

In 1961 the company abandoned intercoastal service even as it tried to renew the PAB Line's operating-differential subsidy. Overall, the maritime business realized over 48 percent of corporate revenues that year but threatened to become a serious financial drain. Managers feared that the maritime books would again flow with red ink if, as required for a subsidy, the company had to spend millions of dollars improving PAB vessels when PAB revenue projections were bleak. Denied the subsidy, the company in 1961 closed all maritime offices except the one in San Francisco, terminated one hundred employees, and put their vessels out to charter.

The last of a once-proud fleet was sold in 1963, ending more than a century of Pope & Talbot maritime history. The proceeds were reinvested in the firm. The responsibility of company mills to fill company ships terminated. The founding families' long identification with the oceans ended, but not their historical interest in international markets. The company's identity now centered exclusively on forest products, and Pope essentially withdrew to his ranch and racehorses. Executive vice president E. N. W. Hunter was effectively given corporate command and ordered to watch out for Pope's interests while Pope spent most of his time away from the firm. Once his sons became proven mangers, Pope intended to pass the helm to them. Quitting the sea proved a key turning point in Pope & Talbot history.

CHAPTER SIX

UNCERTAINTY

1962-1970

INTERIM

FAMILY-OWNED OR -CONTROLLED FIRMS COMPRISE AN ASTONISHING 90 PERCENT OF business enterprises in the United States today. From this overwhelming majority, however, only 30 percent survive in this form into the second generation, 13 percent into the third generation, and a minuscule 3 percent into the fourth. Against the odds Pope & Talbot was again family-owned and -controlled by the late 1940s, but its chances of remaining so were shrinking.

By then "family" had acquired a looser meaning because of various trust arrangements and the growing number of family members. Numerous Walkers, Talbots, and Popes took no role or real interest in the concern. Several wanted to sell stock outside the families. Diversification of their holdings apparently was more important than keeping the firm in the "family." Older family members worried about inheritance taxes.

Between the late 1940s and the early 1950s some of the Walkers and the Talbots sold stock to Blyth & Company, which in turn sold it to the public. Walkers and Talbots, having a declining minority interest and different priorities, generally reduced their ties to the firm. At the last moment, however, George A. Pope Jr. decided to keep his shares, apparently because of his dissatisfaction about the cash return and his concern over the leadership succession. His mother, Edith, brother Kenneth, and sister Emily also retained their shares. Because the publicly held stock had sold widely, no outside investor block arose to question Pope's policies or his continuation as company president.

If George Pope Jr. had been more interested in the lumber business, he might have bought or built another big mill once the Oakridge plant proved a success. But even before maritime operations ended, his interest in maintaining the company was limited to ensuring the leadership succession of young family members. Impressed by vice president Cyrus T. Walker, Pope promoted him to the presidency in 1963 and took the board chairpersonship for himself. At this point Pope controlled approximately 50 percent of the stock, Walker 10 percent, the Talbots 10 percent, and the public 30 percent.

In 1959 the executive office and lumber headquarters had been moved to Portland, closer to the main source of timber. The corporate office remained in San Francisco only because of Pope's California residence. That same year Adolphus Andrews Jr. moved from Portland to the corporate headquarters in San Francisco to concentrate on long-range planning. He sought ways for the company to develop and expand, given its

At Port Gamble in the 1960s tugs and log tows remained familiar sights more than one hundred years after they first appeared.

excellent resource base and high, unused debt capacity. Walker, long interested in property development as well as lumber, began exploiting Washington real estate. In the early 1960s he and Andrews failed to win Pope's assent even to evaluate a big Boeing tract of high-quality timber for a possible bid. In this period, Hillman Lueddemann repeatedly tried and failed to persuade the company to buy more trees to serve the Saint Helens mill. The company drifted, surviving without really prospering during Walker's tenure. In Guy Pope's opinion, Walker was an unaggressive manager.

George Pope Jr. remained true to the faith. He kept the business together for the next generation. He was "passing on the company to you," he told Guy Pope, brothers Peter T. Pope and George Pope III, and Adolphus Andrews. He expected his sons to enter the firm. "We were brought up to believe there was no other choice," Peter Pope recalled. As teenagers the boys worked in the mills and offices during school vacations. In 1959, after college, Guy Pope went to Oakridge, where he did about everything in the mill and woods. The next year he moved to the Portland office. Peter Pope went to business school and then to the San Francisco steamship office. From there he moved to the real estate operation directed from Seattle before moving to Portland four years after Guy Pope. George Pope III mainly worked in Portland, where he handled the raising and buying of timber before moving to Port Gamble. The brothers joined the corporate board during the 1960s. In contrast to their father, they concentrated on business.

Pope & Talbot's long presence in San Francisco ended with Walker's ascendancy to the presidency in 1963. It was a historic change, rued by those who did not want to resettle in the Pacific Northwest. But San Francisco had become redundant, a difficult place from which to exercise any real influence on company affairs. It was a time of change. George Pope Jr.'s pet project, the California manufacture of the fertilizer Fersolin (developed at Oakridge), was terminated after losing money all four years of its production. He was aging, and the abandonment of the maritime trade had turned the firm into a drastically different enterprise. With headquarters and key family figures in Portland, Pope & Talbot became an Oregon, rather than a California, firm. George Pope Jr.'s death in 1978 officially completed the transition.

But before these changes came to pass, a sale of the company was once more seriously considered.

> "The change from one generation to another is always a difficult time for family companies. Pope & Talbot had barely survived the transition from my grandfather's generation to my father's. The shift from my father's generation to the next was carried out with finesse. George A. Pope Jr. sold his stock in the company to his children on credit. Then he and Cyrus T. Walker resigned, making my brother Guy the COO of the company and myself CEO. Thus in one move we gained the responsibility, the authority, and the incentive to make the company more successful."
>
> —Peter T. Pope

George Pope Jr., older executives, and family directors worried about the younger managerial generation's ability to create a rosy corporate future. The firm was making little money at the time, but its liquidation value was very high because of timber holdings. Instead, a plan devised by McKinsey & Company showed the families how to avoid a sale, retain corporate independence, and smooth the transition to the next generation of family leaders. The key to preventing liquidation probably was that the Popes, the Talbots, and the Walkers had enough income from their many valuable noncompany properties and holdings to feel secure about keeping Pope & Talbot intact.

Under the McKinsey terms both the president and board chairperson retired in 1965. George Pope Jr., in a complicated stock arrangement, transferred his portion to his eager children, ensuring them a major voice in the corporation. The way opened for Guy and Peter Pope to take charge and pursue their plans for raising the firm's worth and profits. A new era in company history opened.

BY-PRODUCTS

BETWEEN 1954 AND 1959 PRETAX REVENUE FROM VENEER, particleboard, and chips increased from 1.3 percent to 16.5 percent of the company's wood-product income. Plywood became a prime competitor, cutting deeply into historical markets. The rise of plywood ushered in a new era in wood usage. It was an enormously popular building material. Small amounts also went into truck, airplane, and boat frames; furniture; and luggage. Adopting the latest advances in adhesive chemistry, manufacturers bonded three to five veneer sheets, forming a final product free from the disadvantages of natural wood's linear structure. Competing lumber firms entered the plywood trade in a big way. National output almost doubled between 1960 and 1971.

A decade earlier management had decided that Port Gamble and Oakridge veneer would be more profitable in the long term if the concern also manufactured plywood. In 1961 the firm bought an aging

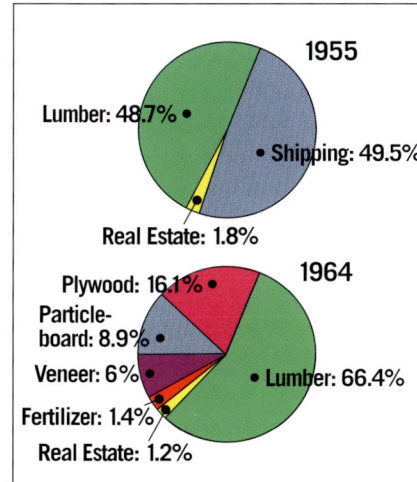

SOURCES OF REVENUE

In ten years, Pope & Talbot's sources of revenue changed materially. Gradual termination of shipping operations, concluded by the sale of the last four ships in 1964, and entry into manufacturing plywood and pressed board were the major factors.

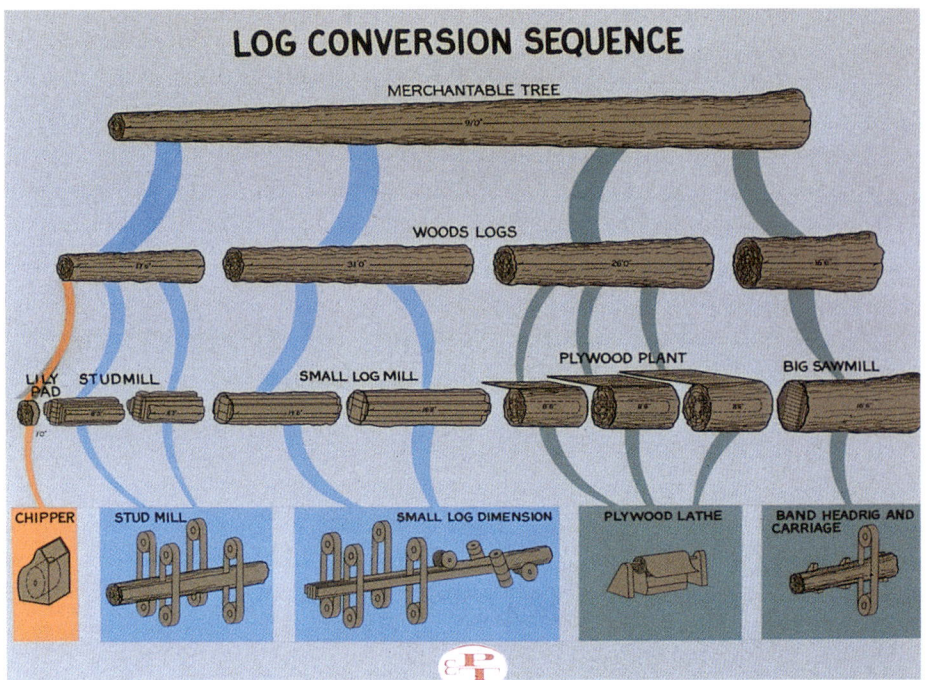

(above) Log conversion sequence.

(above right) A worker stacks plywood that emerges after cutting and planing at the Kalama, Washington, plywood mill in the 1960s. (Don W. Jones, photographer)

plywood plant in Kalama, Washington, across the Columbia River from Saint Helens. The mill was rebuilt, re-equipped, and expanded to improve productivity and lower labor costs.

The Kalama operation's first several years were rocky and unprofitable, especially as more efficient plants came on line. Labor problems also plagued it. Satisfactory results came only when Kalama began emphasizing textured and specialty siding. Aided by a new wave of construction, the plant became the nation's largest producer of exterior fir plywood siding. Management added products designed for home remodelers and installed Chip-n-Saw headrigs to make peeler cores into studs for framing builders. Sales mounted by 75 percent between 1967 and 1968. Plant capacity was expanded—unfortunately, just before the onset of a serious national recession. But over the next few years, "Saint Helens and Kalama basically suffered from a lack of logs," Guy Pope remembered. The Kalama mill became a poor investment and closed in 1974. The newest Pope & Talbot management team had learned a hard lesson about underperforming assets.

Linked to a more diverse product line, resource improvement plans multiplied. Managers reemphasized the sale or use of previously destroyed mill "waste." When feasible, on-site residues were turned into higher-value products. The Saint Helens sawmill was remodeled and upgraded. A line to blow chips to an adjacent Crown Zellerbach pulp installation was added in 1962. After Oakridge's laboratory pioneered the development of medium density fiberboard (MDF) from sawmill waste or veneer, Andrews spearheaded the opening in 1965 of an automated plant there to make both MDF and hardboard and, later, exclusively MDF in greater quantities. Widespread furniture industry demand for MDF, however, inspired fierce competition from newer, more efficient plants exploiting the MDF panel market.

Sharp knives on machines at Oakridge (pictured in 1972) and Kalama literally peeled veneer from logs. Kalama bonded most company-made veneer into plywood.
(Hugh Ackroyd, photographer)

In the postwar era the company began to advertise some of its products to consumers as well as to companies.

BY-PRODUCTS **109**

BRITISH COLUMBIA

GEORGE POPE JR. HAD ENCOURAGED INDEPENDENT THINKING ON THE PART OF THE NEXT generation. He was happy to have his three sons and Andrews direct and otherwise influence company programs. Times were flush. In 1967 the board voted to accelerate internal growth, modernize facilities, and raise output. Marketing and production functions were reorganized and committed to efficient resource utilization. Pope & Talbot Development, Inc., was established to energize development schemes. (This subsidiary was eventually spun off in 1985 as Pope Resources, Inc.) Between 1968 and 1971 the first serious corporate borrowing in two decades secured $23 million for investment purposes.

"Peter, Guy, and I were all excited about the prospect of doing another Oakridge," Andrews remembered about this era. "But the question was where." Rising timber prices could well doom purchasing the big tracts needed for a good-sized mill. He and Guy Pope explored southeastern states in 1968 for timberland and studied southern mill economics. Their findings were not promising. "You have to be at the right place at the right time," Guy Pope recalled, and the Southeast was not it. Soon afterward, Chief Financial Officer George H. Folquet and vice president Clark A. Johnson turned the firm's attention to the tree empire in British Columbia. What ensued was a Canadian expansion that Folquet called "the biggest thing that Pope & Talbot did" during his company career.

British Columbia lay just to the north of the pioneer Puget Sound mill country. For a century B.C. mills vigorously competed with Pope & Talbot. Western mills alternately tried to tax Canadians out of the American market, welcome them into trade associations and price-fixing schemes, and initiate their own operations north of the border. By the mid-1960s sharp Canadian competition forced Port Gamble and Saint Helens to redirect lumber from the East Coast to California and then to Seattle and Portland.

Postwar British Columbia governments promoted resource exploitation and maintaining the viability of small lumbering communities dependent on family-owned mills. The provincial government controlled all but 5 percent of the timber in the province. Stumpage costs were low compared with those in the United

> *"As the new generation of leaders, we made many false starts, but we finally got it right when we purchased the Boundary Forest Products Company sawmills, accompanied by leased cutting rights, in British Columbia. This purchase, like my father's purchase of the Penn Tract and the building of the Oakridge mill, has been instrumental to the company's survival. Without the consistent cash flows, first from Oakridge and then from British Columbia, the company could not have diversified. If we had not diversified, we would have gone the way of other medium-sized public lumber companies. Bohemia Lumber, Brooks Scanlon, Dant & Russell, Medford Corporation, Pacific Lumber—all have gone to the graveyard. Because we diversified into pulp and paper, we are a survivor."*
>
> *—Peter T. Pope*

British Columbia fir larch forest. *(Jon Blaustein, photographer)*

States, and they remained comparatively low. Policies favoring heavy U.S. investment in north-of-the-border pulp and paper manufacturing enjoyed considerable public support in British Columbia at the time. To stabilize employment and communities, great tracts of province-owned land were placed on long-term lease to any firms building or operating mills. The authorities preferred a few large-scale operators to many smaller ones. By 1986 ten U.S. and Canadian concerns were responsible for half the province's wood product output. One of them was Pope & Talbot.

But in the late 1960s Canada was still unknown territory for Pope & Talbot. Surveying possibilities in British Columbia, "a totally unknown area to all of us," Andrews admitted, led the firm to two underfinanced sawmills near the border at Midway and Grand Forks. The company made its largest acquisition in 120 years, paying $9 million in 1969 for a sawmill in each location and for a lumber company at Westbridge. It quickly closed the latter firm but retained its logging rights.

Before this purchase Pope & Talbot had been historically identified with Douglas fir. Extension into British Columbia greatly expanded the various tree species it logged and sold. Multiple species and the cross-border mills redefined the company's traditional production and marketing geography. Meant to serve an exclusively American market, Grand Forks and Midway specialized in two categories of kiln-dried boards and dimension lumber: spruce-pine-fir (SPF) and fir-larch (FL). Fir and larch grew together, usually on lower elevations, and were cut before loggers moved on to the clustered spruce, pine, and fir. Grand Forks also handled another big tree—western red cedar. The red cedar was particularly desirable for such outdoor construction applications as siding, fences, and decks.

Two years of record profits, an anticipated 19 percent increase in housing starts for 1969, and tax advantages figured in the Canadian investment. Long-term considerations, however, were more important: the attractive terms available in British Columbia and the tightening environmental restrictions in the United

States, with their ensuing constrictions on the harvesting and buying of timber. Management saw the end of their American timber supply approaching. Federal timber south of the border was "controlled by the population east of the Rockies," Peter Pope would say, "whereas in British Columbia the public timber is controlled by the province." According to Pope, entering the province "has probably been the single biggest move that has allowed the company to go on."

With the 1969 Canadian acquisitions, Pope & Talbot secured renewable, long-term provincial cutting rights to more than a million acres of timber. By comparison, the firm owned about 130,000 forested acres in Oregon and Washington, some more valuable for housing and other nonindustrial purposes. Although operators were required to reforest in British Columbia, the province imposed fewer and less-costly environmental requirements than in the United States. Operators were not compelled to adopt improved forestry methods until passage of the Forest Act of 1978.

South of the forty-ninth parallel, the political climate had changed considerably since World War II. In the 1960s a broad-based movement had coalesced around the concept of environmental amenities and protection. Increasing numbers of Americans believed the forest was primarily an aesthetic and recreational resource worth protecting from industry. These new attitudes dismissed the traditional conservationism of the big sawmill owners, which entailed a commitment to multiple use of the forests, including their efficiently managed exploitation.

Environmental policies continued to grow in impact. Between 1963 and 1967, Congress enacted federal legislation on clean air and water, wilderness, and endangered species protection. Oregon enforced even stricter state air and water standards, including standards on streams' logging sediment and pulp mill air and water pollutants. Besides passing additional measures to protect the environment, Congress established the Environmental Protection Agency in 1970. The Endangered Species Act of 1973 covered everything above microscopic life, insect pests alone excepted. The various laws threatened to impose major restrictions on the harvesting of trees in the American West. Massive public opposition to the commercial cutting of western trees was gathering on the horizon.

PROPERTY DEVELOPMENT AND HOME BUILDING

CYRUS WALKER'S PRESIDENCY OF THE COMPANY COINCIDED WITH A PERIOD OF POPULAtion growth and rising income in Washington state. Walker himself was very interested in the real estate business. Decades of cautious corporate land-related activities ended. Previously, the Puget Mill division had concentrated on liquidation of cutover tracts near Bremerton, Everett, and Seattle. In 1947 six modest subdivisions had opened in the neighborhood of those communities. Roads, water supplies, and building designs barely exceeded modest standards. Projects were small, and development concentrated on acreage adjacent to improved property. When Washington's population swelled during the Korean War, a further 14,000 acres were scheduled for development.

The real estate division was "a pretty quiet operation" when Peter Pope transferred to its office in 1963. Five years earlier it had begun purchasing development-quality land on Puget Sound. Some of the newer subdivisions sold higher-priced lots and homes than those offered immediately after the war. By 1966 fifteen residential or summer home projects were under way in or near Bangor, Everett, Hansville, Seattle, and

Tacoma. Rolling Green, near Everett, offered six hundred home sites. Locally based sales personnel promoted the properties and arranged financing for builders. Real estate moved from the shadows, contributing 5 percent of total revenue in 1966.

Basically, the company sold off its land on the east side of Puget Sound before starting to develop its land on the west side. Success near Seattle with Shore Woods, a second-home waterfront project featuring a clubhouse, tennis courts, and a swimming pool, prompted the company to stress scenic views and to invest more heavily in amenities. The 1967 corporate growth and expansion plan resulted in establishment of an aggressive real estate subsidiary. Pope & Talbot Development, Inc., built homes in subdivisions in and outside the cities. It dealt in substantially larger parcels of land, buying and selling a thousand or more acres at a time. The value of the realty inventory increased by 34 percent in five years, reaching $4.9 million in 1970.

Rolling Green Estates. *(Don W. Jones, photographer)*

Prime development of the old Port Ludlow property and forests seemed highly feasible. Optimism reigned. The project was based on a twenty-five-year plan for forty thousand people. The 3,000-acre project site was within easy driving distance of Seattle, where it was heavily marketed to sailing enthusiasts. Opened in 1967, the first phase of the recreational and retirement community included a marina, a restaurant, a store, a clubhouse, tennis courts, swimming pools, and eleven hundred building sites for homes and condominiums. The handsome structures blended into rolling hills and forest. The project opening, unfortunately, coincided with a downturn in the economy. By 1969 rising mortgage rates and tightening money markets struck hard at Port Ludlow. The great postwar boom was grinding to a halt, and trouble lay ahead for Pope & Talbot.

> *"When we started, there were so many trees and so few people that the company didn't especially consider owning trees. At the beginning of the twentieth century, it began to dawn on those forest firms with foresight that the timber supply might not be endless, particularly the supply of trees close to waterways. By then, my grandfather was trying to sell the company. The Talbots and the Walkers could not persuade him to expand our forests. My father bought time for the company by buying the Penn Tract. The tract, plus the purchase of public timber, carried Pope & Talbot until environmental mandates started to restrict the cutting of public timber. The growing restrictions would have destroyed Pope & Talbot had it not moved its lumber operations to British Columbia, where the province generally uses its own vast lands for timber production and jobs."*
>
> —Peter T. Pope

DEVELOPING PORT LUDLOW

PORT LUDLOW WAS THE LARGEST PROPERTY development in company history. In the early 1960s the former millsite and tree farm spread along the last beautiful, undeveloped natural harbor on Puget Sound. Pope & Talbot transformed this into a planned community. "It has been very controversial as to how successful it has been," Peter T. Pope has said. "But if you look at the cash flows over the years, they are substantial."

Throughout the West in the 1960s railroads, ranches, and other big property holders were divining the worth of transforming land resources to more valuable ends than growing or transporting things. Pope & Talbot recreated Port Ludlow to capitalize on the popularity of the Northwest's environment, postwar growth and prosperity, and 3,000 acres of corporate land within fifty miles of downtown Seattle. The company also feared that the state of Washington might condemn Port Ludlow for a marine park and pay only for stumpage.

Using its own money, the firm devised a planned community for forty thousand people. The first multimillion-dollar phase of the twenty-five-year plan, partly funded by the sale of the company's last ships, opened in 1967. As a recreational and retirement community, Port Ludlow initially had a maritime orientation. It offered condominiums and eleven hundred residential home sites, financed by installment plans. Structures with a Northwest style blended into rolling hills, forests, and Puget Sound. Pope & Talbot provided a marina, beach club, restaurant and store, model homes, tennis courts, swimming pools, and underground utilities.

The postwar boom unexpectedly ground to a halt about the time that Port Ludlow opened. A fall-off in the Seattle area's economy, rising mortgage rates, and tightening money markets barred any immediate success for the development. Recessions led off 1970 and 1974. It was soon evident that the country was suffering serious economic decline. A decade of inflation, declining wages, and dislocations in industries had started.

Losses forced write-downs of real estate values in 1972 and 1973. The 1973 oil boycott stalled auto-dependent communities at a time when half of Port Ludlow's lots remained unsold. The company borrowed money to broaden the community's appeal. In 1974 and 1975 it added a convention center, championship eighteen-hole golf course, and more condominiums. Marketing of lots around the golf course began. Driven by the economy, lot sales rose, then fell, then rose again in 1977.

A ten-year corporate expansion plan had slated 10 percent of new capital investment for land and community development before nature intervened in 1979. A violent windstorm destroyed part of the Hood Canal bridge on the most direct route between Seattle and Port Ludlow. Sales and usage plummeted, contributing to development losses in 1981. Rising operating costs and managerial problems plagued Port Ludlow. Annual reports tended to pass over it in silence. Bridge replacement in 1982 improved dismal sales and usage without specially rousing hopes. That year's annual report recorded that from a sales perspective, Pope & Talbot was in three businesses now: wood products, timber, and paper.

Throughout 1983 the firm tried and failed to sell Port Ludlow. A year later the state tightened environmental restrictions and placed a moratorium on sewer hookups, stopping homebuilding and sales until a new sewer plant opened. In December 1985, Port Ludlow, when it transferred to the newly formed Pope Resources, was at an impasse. Under George H. Folquet the new company reactivated Port Ludlow and took it in a new direction as a retirement community.

Yachts in the natural harbor reflect the maritime orientation promoted for the new Port Ludlow.

CHAPTER SEVEN

SUCCESSION

1971-1985

NEW LEADERS

IN 1971, GEORGE A. POPE JR. RETIRED AS BOARD CHAIRPERSON AND CYRUS T. WALKER AS president. Pope, then seventy, dictated the succession, even though the company now traded widely on the New York Stock Exchange. After a decade in executive ranks, Peter T. Pope became chairperson and chief executive officer and Guy Pope became president and chief operating officer. Together the brothers implemented the 1967 plan for internal expansion, production increases, and facilities modernization. They increased mill capacities and Canadian cutting rights. With the plant's long-running problem of insufficient timber supply appearing to be insoluble, the Saint Helens sawmill was sold to Boise Cascade in 1971. The last 7,500 acres of nearby forest also went to outside purchasers.

A number of firsts marked the new regime. It made the firm's most costly single improvement since 1853, a complete rebuilding of Oakridge. The reopened facility in 1973 featured a highly computerized sawmill, a stud mill, and an expanded green veneer plant. The sawmill used smaller, low-grade, and defective logs to make chips and other fiber.

The Popes added a $3 million hardwood chip facility at Port Gamble two years later to make the first commercial use of alder. The firm owned one of the Northwest's largest private stocks of this species. The first of several ten-year contracts to supply 132,000 tons a year of alder chips was negotiated with a Japanese pulp firm. Another million-dollar modification enabled Gamble to chip 3-to-6-inch-diameter logs and tree-tops and thinnings.

Pope & Talbot intended to realize top performance. The formula for achieving it was straightforward: obtain not only lower-than-average log costs but also create lower-than-average conversion costs. By 1971 the company spent $5 million to remodel and expand the Midway and Grand Forks mills. They now processed treetops as small as 4 inches in diameter. The mills deployed Canada's first computerized band saw headrigs, which replaced eye-guided circular saws and thus raised the value of logs.

Annual Canadian output increased from 60 to 169 million board feet because of the new capacity for handling second- and third-growth trees once considered too small for processing. As a result, the Midway and Grand Forks plants maintained close to normal output at a time when the Kalama and Oakridge plywood mills temporarily shut down. Future modernizations in British Columbia immensely raised the

(left) Wood chips from the Port Gamble mill are blown into a vessel bound for a Japanese pulp mill in 1972. Chips made a higher-quality pulp than sawdust.

(right) An operator watches a monitor when controlling a machine that simultaneously turned small logs into lumber and woodchips. The machine chipped away the outer part of the log and sawed the inner part, usually into two-by-fours.

board-foot yield from logs and changed the workforce almost entirely to skilled operatives.

Pope & Talbot bought a Hudson, Ontario, stud mill in 1972, primarily to secure its huge licensed timber base in eastern Canada. For the first time production north of the border exceeded the American lumber output. Once Midway and Grand Forks were running full-time, the highly engineered and automated, high-volume Canadian mills supplied half the corporate revenue. All company mills were geared toward maximizing cash flow from each saw log, not just maximizing output.

Canada was a "lucky expansion," Guy Pope said. Start-up problems were minor, and the market turned strong at the right time. "We made money every day." But the firm also made errors. It had expanded Hudson's capacity from 37 to 76 million board feet. Further enlargement was expected, but the Ontario timber proved to be poorer than anticipated. Expanding Hudson to take advantage of its logging rights suddenly made less sense. Pope & Talbot sold the plant in 1974, losing the associated cutting licenses.

The sale was another company first: Management had been speedy in ridding the firm of a potential money-loser. In the past Pope & Talbot had tended to retain unprofitable properties for years, seeking resource supplies for them and waiting for an economic upturn. Concern for employees and communities was more difficult to maintain under the new economic pressures and new company thinking, especially as the corporation moved toward being more widely held by private investors.

Meanwhile, Port Gamble was upgraded, with a new computerized headrig, chipper system, resaws, and edgers. By 1979 annual output had increased by 70 percent. Dry kilns and related equipment enabled the mill to compete in premium markets. Gamble also profited from Alaska's boom in the early 1970s. And until

Nitrogen fertilization was thought to accelerate growth at the Cyrus T. Walker Tree Nursery and Forest Research Center by as much as 35 percent. *(Don W. Jones, photographer)*

NEW LEADERS

barging costs became prohibitive, the mill sold considerable additional lumber around Puget Sound. Smarter use of raw materials, less waste, and better equipment continued to benefit the company's bottom line.

CAPITAL AND TREES

POPE & TALBOT HELD RIGHTS TO OAKRIDGE OLD-GROWTH ORIGINALLY COSTING $2 PER thousand board feet. In contrast, comparable government timber was worth $200 in 1974, and competition for it was often fierce and costly. In 1974, and again in 1982, company executives concerned over high national forest log prices accelerated logging of the old-growth on the Penn Tract. Though advantageous to its owner, this low-cost acreage concealed a larger lesson. In a publicly traded company, timber revenue must justify years of tied-up capital.

If the mills had depended entirely on the firm's 126,000 Northwest acres, Pope & Talbot would have declined in importance. Instead, it continued buying more stumpage than it harvested. By the late 1960s the firm's purchaser role predominated. "We wanted to continue to buy public timber and not tie up all our capital in [growing] trees," Peter Pope explained. Between 1969 and 1978 it secured from 15.6 percent to 42.2 percent, depending on the year, of its logs from state and federal land and from 19.5 percent to 43.6 percent from private owners. In only three years more than half the logs used came from its own property.

Commercial forests continued to rise in monetary worth. Federal timber prices quadrupled on average from 1970 to 1977. According to a 1979 report, the value of Pope & Talbot's trees had increased "at a com-

A forester takes a core sample to estimate the age of a tree on the tree farm. Company foresters also assisted private woodlot owners in growing trees and encouraged them to sell logs to the company.

pound annual rate of 14 percent since 1948." Trees were readily transformed into profits, but were the earnings sufficient to appeal to investors and lenders? The cost of capital, general expenses, and capital gains taxes had to be deducted from gross returns. The forest-products industry remained one of the country's most capital-intensive manufacturing sectors, and much exposed to financial market volatility.

Forest owners preferred to obtain early returns and then to reinvest in a new tree crop, especially under federal tax code provisions existing before 1986. As trees of various ages were removed, fast-growing seedlings could be planted. Tree stand improvement and reforestation were deductible as operating expenses, and money received at harvest was taxed at a 28 percent rate rather than the 46 percent corporate rate. Companies therefore saved as much as 18 percent of their total timber revenues.

For big timber owners the economic maturity of trees consequently took precedence over their productive and biological maturities. Softwood trees, assuming reasonable site and growing conditions, grew at an annual rate of 6 percent. Harvesting them in their prime, firms lost thirty years of peak growth but cut the rotation time almost in half. Instead of paying interest over a longer period of time, they profited on the initial investment and initiated a second round of financing. Shorter rotations lowered the annual volume of output per acre. But the combined volume and profit from the faster harvests exceeded the financial return from longer rotations, which was of key interest to investors in Pope & Talbot.

PUBLIC TIMBER

POPE & TALBOT'S MAIN PROBLEM DURING GEORGE H. FOLQUET'S TENURE, HE RECALLED, revolved around its "ability or inability to buy Forest Service timber on a competitive basis." The volume of timber harvested on federal forests, after a thirty-year rise, fell more than 13 percent between 1972 and 1977. And the average stumpage prices on federal logs climbed 400 percent between 1970 and 1977. A declining flow to mills of federal- and state-owned timber was widely anticipated in the 1970s. The severe environmental restrictions on harvesting of the late 1980s was not.

During the inflationary years between 1977 and 1981, public timber in the United States became an almost irresistible speculation. The highest bidder only paid when harvesting, had ten years to cut the federal timber, and paid no interest in the meantime. Fevered bidding wars for federal and state timber ensued, in which Pope & Talbot vigorously participated. Already higher than appraised values, bid prices continued to rise even when construction nosedived in the United States. The cost of public timber came to exceed both retail and wholesale lumber prices.

Pope & Talbot, as well as its competitors, expected that inflation would further drive up the value of public timber, making contemporary purchases comparative bargains, and that lumber prices would recover before the trees presently under contract would have to be cut. Unexpectedly, the inflationary impulse lost force in the early 1980s, easing the upward trend in lumber prices. In 1984, Pope & Talbot warned stockholders of impending losses should it actually pay for contracted Forest Service stumpage. Sustained losses from the use of this timber would hurt the Oakridge mill in particular.

Defaulting on their contracts, Pacific Northwest timber concerns obtained federal reauctioning of the public timber. Defaulters had to pay the difference between the original and the new bids. Some firms made use of contractual loopholes to avoid liability. Others sold property or closed down. A workforce already

CLEAR-CUTTING DOUGLAS FIR

A PORT GAMBLE EMPLOYEE ASKS, (1) *What is Pope & Talbot's policy on clear-cutting as it relates to conservation; planting of new trees, etc.? Is it correct that some species could not be regrown without the extra sunlight provided by clear cutting? (2) Is it possible for P&T to practice sustained yield forestry; by this I mean that each stand yields a constant amount each year through selective felling and replanting?*

James W. Dyer, chief forester, Oakridge Mills, answers:

(1) We use clear-cutting as a silvicultural tool in order to regenerate even aged stands of Douglas fir similar to nature's past method of clearing the land with fire. The Douglas fir species is intolerant to shade and requires the full sunlight offered by clear-cutting to grow vigorously. . . .

(2) It is possible for P&T to practice sustained yield forestry with uneven aged management (some form of partial cut) or even with aged management (clear cut). Only that amount of stand yield for the year would be harvested somewhere in the forest with a sustained yield program.

The Oregon timber tract [near Oakridge] consists primarily of old-growth timber that has a very slow yield rate. The application of sustained yield management in old-growth timber is not economically practical. Upon the conversion of our old-growth stands to vigorous second-growth, we may be able to approach sustained yield management. Yet, we do apply sustained yield management quite successfully on our Hood Canal tree farms, which are predominately second-growth.

—from "Clearcutting Douglas Fir,"
in Pope & Talbot's *Timberlines* 2 (May 1978)

The company and many competitors practiced sustained-yield management in the postwar era. "Sustained yield" became a forestry slogan, and its meaning was disputed. A realistic sustained yield is the amount of timber a tree farm or licensed cutting area, when worked at a given level of management intensity, can furnish continuously while maintaining a balance between the rate of growth and the rate of logging. In this manner second-growth forests are supposed to come smoothly on stream as old-growth forests (and their mature replacements) decline.

Horticulturists measure the caliper size of seedlings in the Walker Tree Nursery near Port Gamble. By the early 1980s the nursery grew about 3.8 million Douglas fir seedlings.

Plastic tubing protects newly planted seedlings from wildlife in front of an alder stand. The seedlings had been grown for twenty-one months for the company's reforestation program.
(Paul Fusco, photographer)

In coordination with the U.S. Forest Service on cooperatively worked tracts, Pope & Talbot engaged in then-accepted staggered-setting logging and timber utilization practices. That is, it scattered harvested and untouched units of at least the same size across the landscape, and it left as little burnable residue as possible.

Pope & Talbot mainly clear-cut even-aged stands. Clear-cutting in the late 1940s left surrounding forests of mature trees. These were expected to restock the cutover lands through natural seeding. Across western landscapes such "patch logging" usually left moderate-sized harvest units surrounded by mature trees. A clear-cut might be as neat as a mowed hayfield, as Pope & Talbot desired, or too littered with debris and jagged stumps to allow trees to reestablish themselves. To enhance the cutover lands' scenic value, company loggers left trees beside public roads and rivers. Furthermore, they paid attention to watershed protection and prepared campsites and hunting areas for the public.

Pope & Talbot next shifted its reforestation to artificial regeneration methods, which established stands quicker and removed the restriction on cutting adjacent seed blocks. By the mid-1950s it settled into a profitable, low-cost, three-step pattern of clear-cutting, burning, and replanting. Using what some term "augmented yield" methods, company foresters tried to improve on nature, which, by itself, might nurture commercially undesired species or nothing on cutover lands. For new growth foresters variously weeded, planted genetically improved trees, fertilized, and improved fire and predatory protection.

For a few years both tree farms adopted the commercial thinning of thirty- to forty-year-old trees, which paid for itself and accelerated the growth of the remainder. When the lumber market grew strong and prices rose, they shortened the rotation cycle. The patches filled with relatively young and uniform stands containing few snags and little debris and understory vegetation. Large clear-cuts and short rotations enjoyed wide industry support and had no effective challenge before the 1970s.

FALLERS AND BUCKERS

"A TOWERING DOUGLAS FIR LEANS, THEN CRASHES to the ground letting loose a shower of needles and bits of bark. The ground shakes, light suddenly bursts through the half darkness and the sweet smell of fresh sawdust fills the air.... But it's nothing new to the team of cutters, who pause only long enough to allow the tree to come to rest before measuring it into logs and moving on. Experience has made them quick, yet every action is carefully planned. Each tree is unique and unpredictable when separated from its roots and deserving of the loggers' undivided attention.

By noon the cutters are gone, and so is the cool shade. The delicate rhododendrons that thrived in the shadows for half a century have already begun to die in the bright sunshine, not to return until the third generation of fir provides enough shade to protect them. What was yesterday a stand of mature second growth timber is now a jumble of debris three feet deep.

The methods used by Pope & Talbot loggers to get trees to the sawmill vary—but whether it's high lead logging of old growth Douglas fir in western Oregon, cat [Caterpillar machine] logging second growth fir on Washington's Olympic peninsula, or selective cutting in the piney woods of the Boundary in British Columbia, it's all the working place of loggers....

Hard hats, chainsaws and sophisticated loading equipment with airconditioned cabs may have dimmed the logger's rugged reputation, but the fallers, buckers, yarders, chokersetters, hooktenders and bull buckers who hold their safety meetings in a 'crummy' (it's just a bus) still don't see themselves as typical modern industrial workers....

Though he may argue the point passionately, today's logger isn't much different from other workers. He has a family, a mortgage and is likely to work many years for the same employer. Logging camps remain only in remote areas....

There is only one way in which the logger continues to stand apart from other workers. While considerably better off than his 19th century contemporary, he has still chosen a very dangerous occupation. According to a NIOSH study (National Institute for Occupational Safety and Health), a logger is 25 times more likely to die as a result of job injuries than other workers, and for each of the 300,000 loggers in the U.S., one work day a year is lost due to a non-fatal injury.

Pope & Talbot loggers, however, have successfully avoided becoming a part of these scary statistics. Safety records for all three company areas... are not only well below industry averages, but below what experts project for a company of our size....

Implementation of the [safety] program is left to local managers and loggers themselves. Safety help comes from the outside, too, from the legions of safety watchers in the U.S. and Canada, as well as from miles of words written on logging methods and safety....

Ironically, many of the injuries all too common today were unheard of a hundred years ago. Logging trucks and heavy motorized logging equipment brought traffic accidents to the bush; and chainsaws, which were perfected in the Northwest, added a new injury category to worry safety directors—chainsaw cuts. The noise levels reached by logging equipment are also a hazard to the hearing.

It didn't take the timber industry long, however, to rise to these challenges. Ear plugs or muffs and leg pads are now required protective equipment, and seat belts are provided, though only required by law in Canada.... As Port Gamble forester Milt Philbrook points out, 'There are old loggers, and there are dumb loggers—but there are no old, dumb loggers.'"

—from "Fallers, Buckers, Chokersetters, and Chasers," in Pope & Talbot's *Timberlines* 2 (fall 1978)

In the 1980s and 1990s Pope & Talbot continued to adopt advanced logging technologies because of where and how the company might or had to cut trees. It entered the ponderosa pine country of South Dakota and Wyoming. While its working forests around Puget Sound remained important, it quit logging on Oregon's Willamette National Forest. Logging rights in British Columbia's interior grew to more than 2 million acres. So the company cut timber over

A faller-buncher clears low-value alder for Port Gamble to convert into hardwood chips for pulp mills. The cleared land would be replanted with high-value Douglas fir.

larger and more physically diverse areas, and in jurisdictions where regulations both varied and intensified.

Alterations in market demands, technology, and environmental sensitivity modified traditional logging practices and introduced new ones. Fierce competition and volatility in the industry encouraged innovation. Aerial and satellite photography helped plan cuts. Computers quickly organized logging "shows." Maintenance trucks equipped with tools and spare parts made on-the-spot repairs. To help keep the struggling Port Gamble mill open in the late 1980s, local employees designed a small self-loader. It reached otherwise inaccessible private woodlots. A mounted, mechanical arm loaded logs onto the one-of-a-kind truck.

In British Columbia during the past three decades, "We went from the conventional logging system of hand-felling and hand-bucking to faller bunchers, grapple skidders, and log processors," David Doumont, a Pope & Talbot Ltd. foreman told Nelson, B.C.'s *Venture* magazine in May 1999. "Nobody touches a stick." The self-propelled faller-buncher's hydraulic apparatus sheared off a tree near the ground and used its large claws to pile six or eight fallen trees together. Some models stripped limbs. Faller-bunchers worked best on small to medium-sized trees on relatively level ground. Rubber-tired grapple skidders dragged the bundled trees to landings in the woods. There, log processors cut off the limbs and cut logs to lengths that trucks could carry. A "butt 'n' top" loader picked up the logs and put them on a truck. Upwards of one hundred logs filled one of his trucks, not the half-dozen medium-sized ones once available.

Heavy equipment operators sat in cabs over control panels. Crew size had shrunk. Cable rigging and hand-run chain saws were usable only at higher elevations impassible to heavy machinery. New equipment made timber buckers, fallers, and chokersetters largely obsolete. But the logger's art would never cease to be needed. Knowledge and experience counted in setting up, repairing, and handling intricate equipment. Each block still had to be treated in a different way. Individuals had to gauge terrain, water-flow patterns, and other variables.

To lessen logging's environmental impact, lighter and stronger machinery and rubber-wheeled equipment was used instead of steel-threaded equipment. In the late 1980s helicopters logged steep slopes that tractored devices harmed or could not reach. In South Dakota the company occasionally contracted for horse skidding, an economical practice in which horses maneuvered logs through thick ponderosa pine stands, avoiding big tire tracks and scarred trees.

The new machines and methods reduced the impact on the land and made for handsomer and more productive sites. Skidders and front-end loaders before the 1970s needed large forest landings. The machines packed the soil almost as solid as cement. Little grew back very well. A crew using modern skidders, processors, and loaders needed only a small landing area, and they better preserved the ground for new growth. If clear-cutting, the crews tried to leave trees protecting stream corridors and wildlife habitats. And they reentered salvaged cuts to soften the edges, growing new timber and recontouring them to fit in with the shape of nearby mountains.

Since the 1960s, enhanced forestry regulation by governments in the United States and Canada and the growing environmental protest movements affected everybody's logging. "You've got to re-do your thinking to cope with the changes in forest policies," Doumont noted. "If we see that we're making ruts in the bush, we've got to stop work." Crews risked fines for discarding small oil cans in streams or lakes. And improved equipment "has allowed for a lot cleaner, quicker road building," thus dampening a major flash point for protesters.

Heavy mechanization of logging began to cost loggers jobs in the 1960s and 1970s. Job losses in logging (and sawmills) worsened in the 1980s. Yet the same advanced technology—coupled with intensified safety training and family-assistance programs—notably improved crew safety. Accidents still happen. Only commercial fishing is a more hazardous occupation than timber cutting in the United States, according to 1999 federal figures. But in 1998 the company's Canadian loggers suffered no time-loss incident. It was a remarkable achievement in so dangerous an occupation.

Worked by a tug, logs are loaded in Port Gamble for the high-priced export market. Privately owned logs had grown more valuable at home and abroad as public forests were placed off-limits to loggers.

hard-hit by mill closures faced more layoffs. Contract holders sought relief. The Forest Service initially granted a two-year moratorium. Approved in 1984, timber relief legislation allowed operators to abandon a portion of the 12 billion board feet in outstanding timber contracts without penalty. Otherwise, Pope & Talbot and many of its competitors would technically have been broke.

Pope & Talbot returned the maximum 200 million board feet allowed each contractor. Required to harvest 85 million board feet, it faced some $7 million in penalties on tracts it could not afford to cut. In a final settlement in 1990, the company transferred 6,520 acres, worth several million dollars, in the Penn Tract to the U.S. Forest Service. Eight timber sales were also returned to the state of Washington. The era of timber contract speculation ended, for good and substantial reasons.

DIVERSIFICATION

PETER POPE LOOKED AT BUSINESS IN BROAD STRATEGIC TERMS. INFLUENCED BY BUSINESS cycle doctrine, he directed examination of industries whose alternating cycles of recession and recovery balanced those of lumbering. Managers studied how Pope & Talbot could obtain a broader and more diversified earning base. The company advertised itself in 1971 as "actively seeking acquisitions in related fields countercyclical to our present operations." A year later it spoke explicitly of diversifying outside wood products. Management even considered entering the garbage trade before deciding that it should diversify into an area closer to its historical experience.

The squeeze was on during the global economic upheaval of the 1970s. American companies struggled to find better means of coping and prospering, not disappearing. Cutting-edge business thought emphasized that efficiency, adaptability, resourcefulness, and specialization were mandatory in the trying times. Manufacturers must introduce still newer and more varied and upgraded products and adopt still better administrative methods. Mergers, acquisitions, decentralization, production shifts offshore, and retrenchment through corporate downsizing were among the strategies adopted in boardrooms and executive suites. At various times Pope & Talbot pursued all but the merger option.

A larger and more sophisticated entity emerged in 1977. At that time Pope & Talbot owned four sawmills and a declining timber base. Business and building cycles and seasonal variations, like the annual spring pickup in lumber orders, made it a historically volatile corporation. Relatively weak housing markets, from the flattening out of the postwar baby boom, foretold worse difficulty during the coming decades. Management concluded that a pure lumbering firm could never maintain rank as a viable public corporation. Despite the troubled times and problematic future, revenues of $160 million made 1978 a record year. Assets increased by 23 percent over 1977. Dividends had risen for eight years in a row. In 1978 the corporation launched "the most ambitious expansion program in Pope & Talbot's long history." The ten-year plan called for significant investment in several sectors to make it a more diversified company.

Due to illness, Guy Pope left the firm in 1978. William A. Whelan, a veteran forest products executive, replaced him as president and chief operating officer. As chief of the Western Wood Products Association, Whelan would be significant in winning the federal reauctioning of public timber. His ascension marked an important transition for an increasingly public family company. Henceforth, professional managers, rather than Pope heirs, usually occupied the presidency.

A spate of heavy borrowing, $15.8 million altogether for new projects and debt refinancing, reflected the makeover. Long-term debt doubled in 1978, to $24 million. The ten-year plan assigned 40 percent of future capital spending to pulp and paper, 30 percent to solid wood products, 20 percent to resource expansion, and 10 percent to land and property development. In 1979, Pope & Talbot reincorporated in Delaware, with an increase in authorized common stock from 5.5 million to 8 million shares.

The decision to invest in a non–housing-related industry (pulp making) was a major step in Pope & Talbot's transformation from a purely wood products firm. Its first diversification since the abandonment of

> *"As time marched on, it became obvious that Pope & Talbot had to gain more of a public following. The founding families had grown to the point that no single person among them owned enough stock in the company to influence decisions. The descendants of the founders had also been selling stock, thus increasing public ownership. The company of the future would have to be a public company with family characteristics rather than a family company with public characteristics. Diversification was necessary to enable the company to have more stable earnings and to service more debt so it could grow. In diversifying, we followed the route of our industry into pulp and paper."*
>
> —*Peter T. Pope*

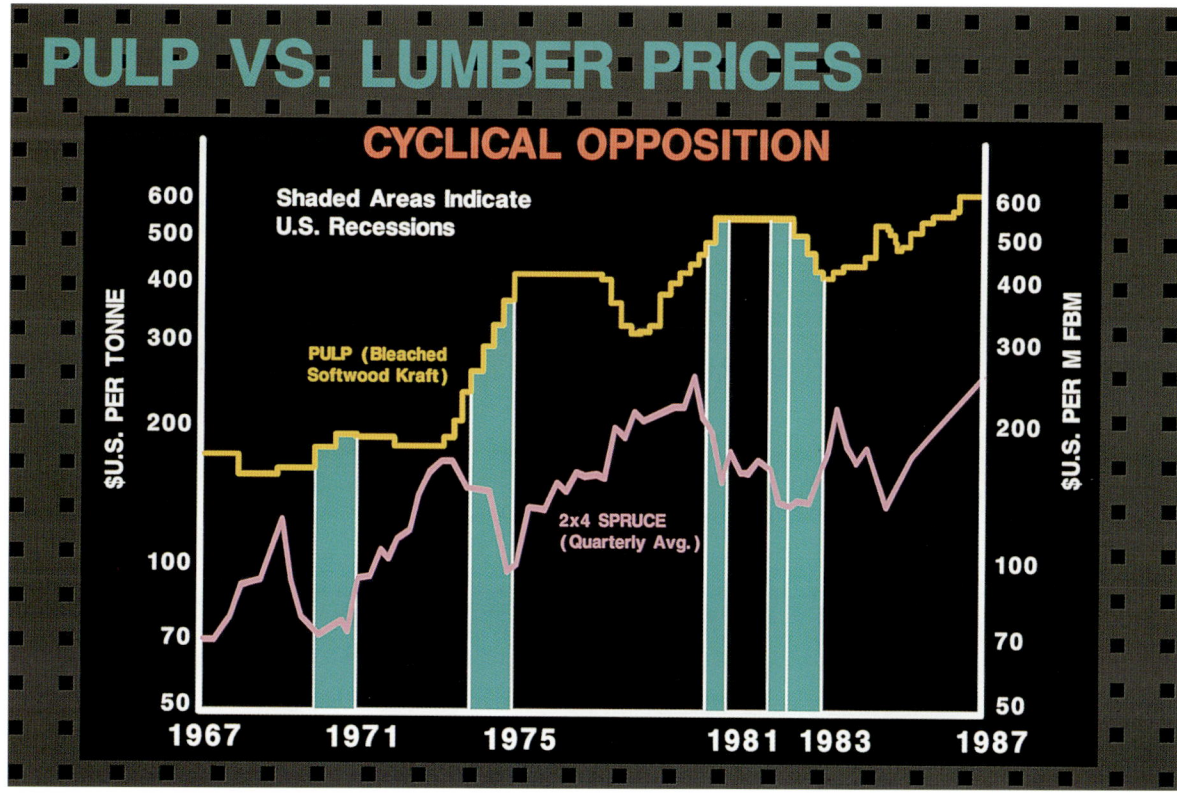

Pulp–lumber cycle analyses like this one moved the company into pulp making.

ships would be followed, two years later, by another in absorbent paper products. Diversification involved offering a complementary mix of products, and projected stability and growth. Pulp looked very promising for many reasons. Pulp demand had swelled. In 1952 pulp and paper profitability had surpassed sawmill profitability in the United States. At the state level the value of Washington pulp exceeded that of lumber in 1960. Pulp mills generated more revenue and profit than any other segment of the forest products field.

Buying into pulp fit into a well-established industry trend, too. Beginning in the 1950s, lumber and plywood producers heavily invested in pulp and paper. Companies in the latter industry cultivated direct connections with lumber producers. Worldwide, a third of the leading one hundred pulp and paper concerns disappeared as independent entities between 1974 and 1990. Many of the remaining ones dramatically increased in size. Regional outfits became national corporations.

In 1978, American Can Company sold to chip- and sawdust-rich Pope & Talbot half-interest in its state-of-the-art pulp mill in Halsey, Oregon. Its outsized machinery made the pulp slurry that was then transferred to American Can's adjacent paper mill. The rapid closure of sawmills in the region had reduced Halsey's supply of sawdust and chips. The Oakridge sawmill, only sixty-five miles away, potentially guaranteed the vital supply at fair-market value. Because Pope & Talbot knew little of pulp making when it entered the business, American Can (later, James River papermill) directly supervised the joint venture until 1989, when the employees transferred to the Pope & Talbot payroll.

Oakridge was now economically integrated into pulp making. "Economic integration means the balance we maintain between the products we produce and the products we consume," the 1990 annual report

Bales of wastepaper are sorted before de-inking, pulping, and turning into new paper at the Eau Claire plant. *(Paul Fusco, photographer)*

On Eau Claire's backstands 175 rolls are unwound, folded, and combined into a continuous web of facial tissues. The web will be cut into individual bundles, packaged, and encased.

explained. In a related move, the company adopted the business unit form of organization, dividing itself into Wood Products and Pulp and Paper divisions.

In diversifying, the managerial strategy was to level off the inherent and fairly regular business cycles typical of the wood product industry. Pulp cycles, both in demand and in prices, typically rose and fell in multiyear blocks. Historically, pulp prices usually ran counter to the rise and fall of building material prices. Pulp tended to peak with the high point of the business cycle. Its market weakened directly as the economy weakened, when demand for paper fell off. But demand for construction lumber generally lagged behind the business cycle. Lumber tended to peak before the business cycle peaked and to recover later in the recession, when lower interest rates and government programs stimulated housing. Although the lumber market was strong in 1978, experts predicted that demand would level off during the 1980s. Pulp and paper, though, would enjoy a 3 to 5 percent growth rate, assuming that predictions became reality.

Contrary to conventional business thinking, Pope & Talbot added to its assets at the low point in a commodity's cycle. "We are basically commodity speculators," Peter Pope explained. "We do it through the acquisition of sawmills or pulp mills, but we are basically speculating in commodities." Pulp demand and revenues were low in 1978. Paying $24.4 million for half of a modern 120,000-ton plant at Halsey therefore appeared to be a very good price.

A corporation that speculates, Pope clarified, needs to follow several rules: "Not too much debt, as you are going to have some really bad times that you won't survive." Furthermore, "You've got to have a good sense of commodity markets," the timing of fluctuations, and the company's relationship to these fluctuations. You must be able to see where trends are headed "and take advantage of them or get out of the way." And "you have to have a tremendous amount of courage to take the plunge when nobody else does." That is, buy when everybody says things will get worse, not when everybody says things are improving. After a year or two, according to the formula, acquired assets will increase in value.

Family pressure also encouraged product diversification. Members of the Pope family wanted to broaden their stock portfolios. They therefore wished to sell company stock and reinvest in other enterprises. As an alternative, Pope & Talbot diversified the corporation's product offering, a move that forestalled family sell-offs. Diversification also provided, at least in theory, additional opportunity for family involvement in the running of the firm and its growing number of components.

In 1980, Pope & Talbot expanded into another complementary area: to make private-label absorbent paper products under store-chain labels. Using its own pulp and waste paper pulped in its own tissue and (soon to come) diaper plants promised more economic integration. It was also a plus that the absorbent paper industry was "more stable than lumber and is the least cyclical of the paper markets," the 1980 annual report noted. It had a countercyclical virtue, for its up-and-down cycles differed from those of pulp. Pulp demand fell during recessions. Lower-priced, private-brand absorbent products sold best, and made retailers more money, when consumers were otherwise trying to save money during recessions. In theory, Pope & Talbot managers expected to succeed. But it was "a whole new ball game," Guy Pope told the author. "We didn't know the industry." It "was hard to realize that at the time."

Two undercapitalized and partly outmoded Wisconsin paper mills in Eau Claire and Ladysmith were purchased for $35 million; industry veterans were deployed to run operations and sales. Demand for both generic and private-label absorbent paper seemed to be heading upward in 1980, even as the national economy faltered. The tissue specialty market, Peter Pope told analysts in May 1990, was "the least capital-intensive of all the paper products business." It was also, he later said, "the only sector of the paper industry we could afford." From a size standpoint, therefore, Pope & Talbot could aspire to become the "dominant" manufacturer. The goal of dominating a particular market niche was an entirely new one for the company. No thought had ever been given to achieving a similar status for its lumber or pulp operations.

Deploying an annual capacity of between 65,000 and 70,000 tons, the new owner immediately became the nation's third-largest producer of generic and private-label tissue products. After recycling waste paper at Eau Claire, the two mills converted the paper thus produced into bathroom and facial tissue, towels, napkins, and disposable diapers. The concept of a high-quality disposable diaper was just emerging, and it seemed feasible to upgrade the diaper plant to produce for this new market. Because of transportation costs, the plants sold only to retailers in the Midwest and the East. Pope & Talbot at once spent $16 million modernizing the plants. It developed means to use cheaper waste paper, upgrade water treatment, and improve shipping and customer service. Efficiency and output climbed, diaper sales increasing by 35 percent between 1983 and 1984. Another $8 million was invested in the plants in 1984, despite industry overcapacity.

Once it became experienced in tissue-making, Pope & Talbot, pursuing economic integration, intended to further integrate the Wisconsin and Oregon operations and install a tissue machine at the Halsey pulp mill for on-the-spot pulp use. Instead, the firm bought the remaining outside half-interest in the pulp mill in 1983, at a price of $14 million. The next year it reoriented the plant to the manufacture of wet lap pulp for regional buyers, mainly newsprint makers. Until then, two-thirds of the mill slurry had been sent to the single nearby paper mill, now owned by James River.

Tissue-making profits were satisfactory until 1982, when the rival Fort Howard Company expanded its capacity. Under the pressure of huge investment and production, it slashed prices and did anything and everything to keep its mills running. In response, Pope & Talbot altered its tissue product mix. The mills converted their industrial tissue machines to higher-profit consumer tissues. Disposable diaper manufacturing was also reconfigured from using waste paper to using fluff pulp, which Pope & Talbot did not make. And elastic leg openings and other diaper modifications were made. Kimberly Clark, Weyerhaeuser, and Procter & Gamble, meanwhile, mounted additional competition in all lines. They were "way ahead of us on the marketing," Guy Pope remembered. Though the company tried to catch up, Pope & Talbot's experience in a wood product industry that historically did relatively little marketing counted against it. "We had a tiger by the tail" in the diaper market, Andrews told the author. "We never built the management strengths that we needed. The concept was sound but the execution lacked something."

THE POWER OF THE ECONOMY

POPE & TALBOT EXPERIENCED SEVERE STRAINS ONCE AMERICA'S POSTWAR EXPANSION started to grind to a permanent halt between 1965 and 1973. Economically advanced nations were returning in strength to global arenas. Foreign producers increasingly offered better prices or products than Americans. A global economy was fitfully emerging—in wood and other primary products initially in the 1960s—that challenged the United States's command of huge domestic and foreign markets. Asia's large purchases of West Coast logs translated into higher federal timber prices for the company and its rivals, for instance. Pope & Talbot reacted in customary fashion in 1966, when a sharp recession struck its industry: It cut back

Pope & Talbot manufactured "Pert," "Teddy Bear," and other store-brand tissue products during the 1980s. *(Paul Fusco, photographer)*

production and laid off employees outside its core operations. As international trends speeded up in ensuing decades, traditional company ways were severely tested.

In the 1970s food shortages, an oil crisis, double-digit inflation, and recessions early in the decade brought about an economic trauma more severe than anything the country had seen since the Great Depression. Basic American industries closed or shrank, mass layoffs spread, productivity and nonfarm income plummeted—all stripping away the illusion of security. Recessions headed up 1970 and 1974, and two more followed in the early 1980s. Sales of Port Ludlow lots stalled during the 1973–74 oil boycott. Inflation and unemployment struck simultaneously between 1975 and 1980.

The Northwest's forest product industry reeled from a combination of strong southeastern competition, rising Canadian competition, and skyrocketing home prices that doubled the cost of a new single-family home. Canadian lumber had 19 percent of the American market in 1975; by 1985 it was 33 percent. After the recessionary 1974–75 period, lumber demand and retail prices hit peak levels in 1977. Then inflation climbed to 8 percent in 1978 and a staggering 13.3 percent in 1979; home mortgages reached double digits.

New home starts collapsed in the summer of 1979, causing an epidemic of mill closures. Company disappearances and restructurings hastened closures. Between 1977 and 1991 virtually all small to medium-sized Northwest lumber companies were acquired or went out of business. Giants like Georgia-Pacific, Weyerhaeuser, and Champion International likewise closed mills. Between 1976 and 1983, Oregon lost forty-nine sawmills, many of them in small, single-industry towns that had prospered in the postwar era. Many other sawmills shut for weeks or months to shave costs. Pope & Talbot closed the Oakridge mill temporarily in 1980; after reopening, there were fewer weeks of full operation each year.

The company's 1981 annual report described the industry's depression as "reminiscent of the 1930s." Federal Reserve Board interest rates, which had been raised aggressively between November 1980 and August

From on-line diagrams in the control room, a machine operator monitors washing and the oxygen delignification process, installed in 1993, at the Halsey Pulp Mill.

1982, further restrained Pope & Talbot's construction and property markets. The recession between July 1981 and November 1982 forced Americans to confront their worst economic circumstances since the 1930s. In rough financial seas the company stopped its capital expenditure program and essentially devoted all cash flow to retiring corporate debt. It reduced long-term debt from $74 million to $53 million by the end of 1982, achieving a low (for its industry) 35 percent debt-to-capital ratio. Every company facility operated intermittently in that year.

Austerity reigned. Executives urged everyone to watch cash flows, minimize supply purchases and all expenses, and be aware of the terrible situation. To compensate for still falling prices in 1983 and 1984, Pope & Talbot's sawmills pushed for great efficiency and volume and higher-valued products. Absorbent product managers looked for additions or changes offering the company countercyclical protection. The firm's commitment to invest in productivity and innovation bumped hard against toughened financial controls and targets. New choices were ahead, some that would yield wrenching consequences.

ENTRY INTO PULP MAKING

POPE & TALBOT BOUGHT HALF-INTEREST IN A STATE-OF-THE-ART KRAFT PULP MILL IN Halsey, Oregon, in 1978. (It acquired the other half in 1983.) For several years a paper mill on the Halsey site purchased Halsey's pulp and managed the plant while Pope & Talbot learned the pulp business. Experience with Halsey subsequently encouraged the company to become a tissue maker in 1980 and to buy a huge pulp mill, Harmac Pacific, in British Columbia in the late 1990s. The Halsey mill, originally opened in 1969, proved a demanding but ultimately beneficial facility for Pope & Talbot.

Ever since the industry's scientific renaissance in the mid-1950s and early 1960s, pulp mills have required frequent and expensive equipment and manufacturing process changes. By the 1980s American pulp

HALSEY PULP AND THE ENVIRONMENT

PULP MANUFACTURING IS STRAIGHTFORWARD IF technically complex. All paper is made by first converting wood, rags, or other fibrous material into a refined slurry. In the widely popular kraft mill, a digester "cooks" cellulose sawdust or wood chip fibers in an alkaline solution under heat and steam pressure. Cooking dissolves the glue-like lignin binder. The fibers separate as the cooked materials are blown into a blow tank, from which the slurry is pumped over coarse screens to remove knots and partially cooked wood. Washing removes the dissolved lignin and other impurities. These are concentrated and burned, along with the spent cooking liquor, to recover energy and chemicals. Several bleaching stages follow at Halsey, using bleach manufactured on the spot.

Before Pope & Talbot entered this industry, pulp makers had already begun focusing on toxicity and higher levels of air and water purity, and sought optimum locations for environmental as well as commercial reasons. Markets, competitors, and government regulations kept raising the stakes.

Pollution control combines reduction at the source with proper treatment and dispersion. At its 1969 opening, Halsey was considered the best environmentally designed paper and pulp mill in the Pacific Northwest. It lacked the once-common dangerous waterway discharges, darkly belching smokestacks, and noxious smells of classic paper mills. To preserve the Willamette Greenway, the plant was erected 2.5 miles from the Willamette River. To collect and destroy odorous gases, it had the first low-odor recovery boiler in North America. The typical sulfur smell was absent.

From cooking the chips or sawdust through cleaning the pulp, water is second only to wood sources in importance. Pulp mills must be near rivers or oceans for supplies and water discharges. Halsey opened with state-of-the-art primary and secondary wastewater treatment systems. Sedimentation reservoirs, screens, centrifuges, and thickening tanks removed most of the total suspended solids from river-bound effluents. Aeration and nutrients in two large effluent ponds reduced biochemical oxygen demand. The ponds promoted growth of organisms that "ate" pollutants before final discharge.

However, manufacturing processes and atmospheric conditions together still allowed some undesirable matter to escape along with harmless water vapor from the smokestack's plume.

By the 1970s Pope & Talbot was thinking "green" everywhere it operated. An optimistic faith in the power of environmental engineering and a determination to stay competitive and meet or exceed regulations powered the thinking. "We remain committed to finding responsible solutions to environmental problems as we encounter them," stated the 1989 annual report.

In 1987, Halsey upgraded its air equipment and controls to new state-of-the-art standards. With a third digester and a consequent 40 percent capacity increase, the old equipment would have worsened air quality. Improving the recovery boiler's collection and cleaning of flue gas resulted in today's 99.5 percent air particulate removal. Upgraded scrubbers removed 99 percent of limekiln dust. (Recycling pulping chemicals saves money and reduces the environmental impact.) Since then, it has improved collection and incineration to prevent release of other contaminated gases. Expanded scrubber systems in the bleach plant have further reduced air contaminants.

With environmental improvements between 1989 and 1995, Halsey labored to trim pollution output to the lowest of any American bleached pulp mill. In 1990, the Halsey mill was the first chlorine user in Oregon to lessen river discharge of dioxins, unwanted by-products of chemical reactions. It also agreed to install the best technology available for pulp bleaching. In association with Oregon State University, the mill began testing naturally safe bioremediation. So it created one of the country's largest experimental wetlands projects for secondary water treatment. And it initiated and helped pay for the Willamette River Study, the most comprehensive assessment of the river's health to this day.

Company vice president Bill Frohnmayer sent a scientific team around the globe to identify the best technologies. "To environmentalists, he has taken dioxin seriously, searching for ways to eliminate it rather than just meeting regulations,"

Portland's *Business Journal* noted on June 8, 1992. "To some in industry, he has taken [Pope & Talbot] so far ahead of competitors in environmental control that other mill operators look bad by comparison."

The mill searched for a workable, totally chlorine-free (TCF) technology. As an interim step, the firm spent $25 million in 1993 to install an oxygen delignification system. Three new concentrators on the recovery boiler raised the amount of solids in liquid and improved air quality by keeping flue gases from coming into contact with odor-causing "black liquor."

A new oxygen system, measured by river discharges, reduced dioxins by 40 percent, effluent color by 30 percent (to look like apple juice), and certain absorbent compounds like chloroform by 30 percent. Subsequent changes further lightened the wastewater color. Halsey complied early with all state environmental regulations.

In 1996, Halsey rebuilt the bleach plant to temporarily use an elemental chlorine-free (ECF) process. Effluent from ECF and TCF mills have the same measurable ecological impact on rivers. Switching to ECF processes preceded by five years the federal requirement to eliminate dioxin. Today, dioxins are no longer detectable in the mill's process water or final wastewater.

The mill started another big environmental overhaul in 1998. It planned to spend an estimated $35 million for new equipment and engineering by the end of 2000 to meet stricter Clean Air and Clean Water rules. These rules dictated that kraft pulp must be bleached without the use of elemental chlorine, and that the average chlorinated organics in effluent must fall. Because of past actions, only modest changes in pulp processing were needed for Halsey to meet and exceed these higher air pollution standards.

The new process uses chlorine dioxide, which must be made on site, as the primary bleaching agent. It is more environmentally friendly to the wastewater. Improved pulp washing before bleaching will also reduce the amount used of this vital chemical, lower operating costs, and further improve water effluent. Federal standards climb again for 2006, suggesting additional air quality improvements in the years to come.

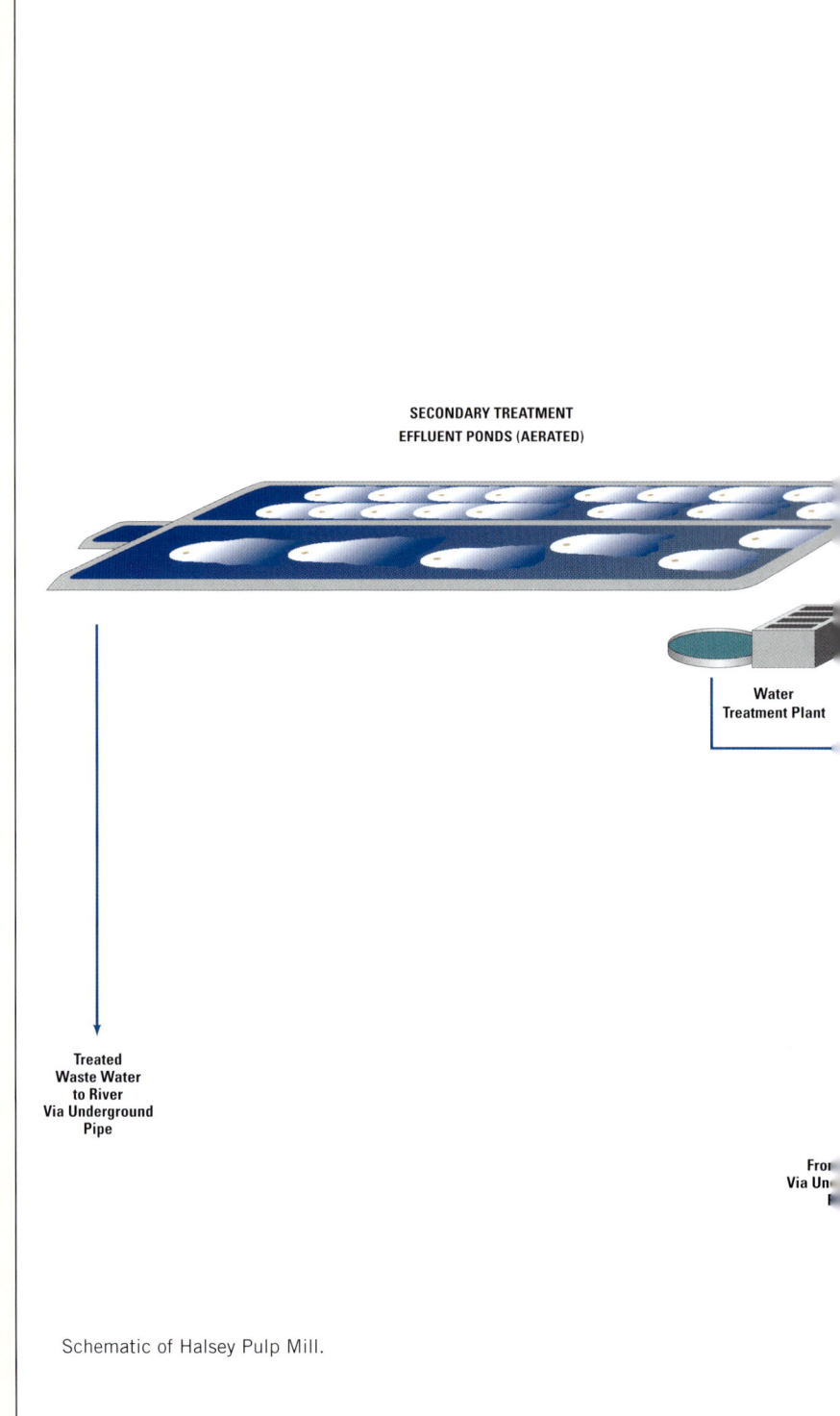

Schematic of Halsey Pulp Mill.

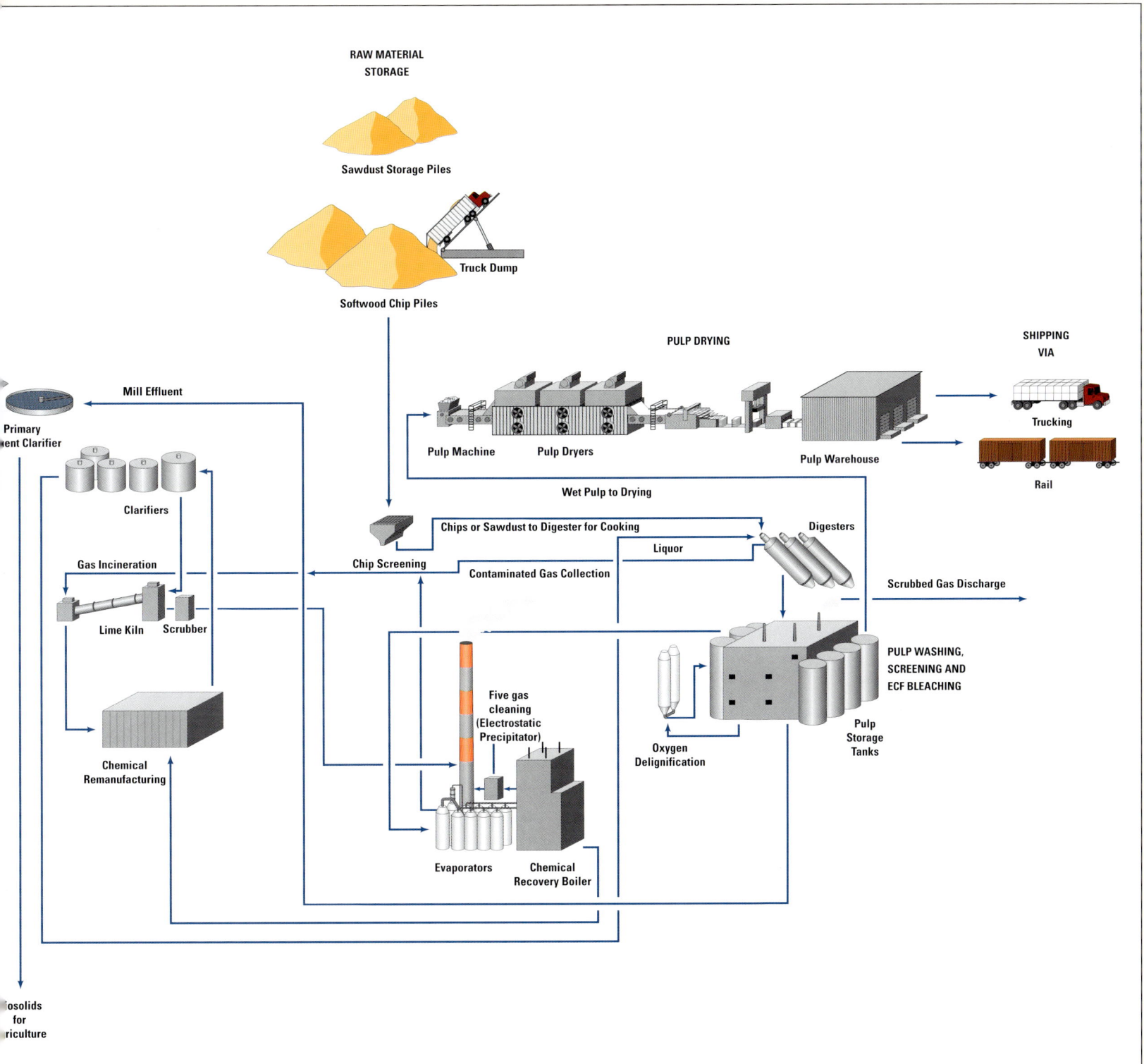

Halsey Pulp Mill in the 1980s.

makers have had the additional burdens of meeting growing government regulations and responding to international competition. The huge costs necessary, along with other factors, meant that additional pulp mill capacity would be achieved by modifying existing plants, not building new ones. Halsey proved to be the last pulp mill built west of the Rocky Mountains.

During the 1980s, at a time when American manufacturers in general were shrinking factory and machinery investments, Halsey was being redesigned and transformed into a bleached chemical market operation. It went from converting chips and sawdust to slurry (for the paper mill) to producing "market" pulp, a basic international commodity. Today sawdust accounts for 60 percent, and wood chips for 40 percent, of its pulp.

Halsey also installed new drying equipment to save energy and raise profits. The action improved pulp strength and reduced pulp weight and thus shipping costs. The geographic range of customers could thus be extended. The resulting White Gold 49, semidry sheets of pulp, won a strong West Coast niche market: newsprint makers. For several years the plant also made some hardwood-based pulp, after extending its supply base into Montana and eastern Washington. In 1987, Halsey installed a third digester. This $8 million unit raised annual capacity by 38 percent to its present 180,000 metric tons. By the early 1990s, the mill probably had one of the highest rates of return among western American pulp mills.

Previous expenditures paled against the $70 million that Pope & Talbot spent upgrading and adding to the facility in 1989–94. (Environmental protection upgrades accounted for $30 million of the sum.) Improvements included a $4 million computerized process-control system. The $34 million spent for the new pulp dryer and a new bleaching system constituted the largest single-plant capital investment in company

138 SUCCESSION

Halsey's "White Gold 49" pulp is trucked to West Coast newsprint manufacturers in the mid-1980s. *(Bruce Forster, photographer)*

history, except for acquisitions. Very heavy spending continued in 1998 and 1999. Some $35 million in new capital projects will simultaneously bring Halsey into compliance with stricter Environmental Protection Agency requirements and improve pulp quality.

CORE SECTORS

POPE & TALBOT PAID CLOSE ATTENTION TO CORE SECTORS DURING PRODUCT DIVERSIFIcation. Half the planned capital spending was allocated to wood products. Although the bulk of this spending was in Canada, millions of dollars remained in the United States to buy and modify sawmills and purchase timberland. Under Whelan's guidance the American and Canadian mills, Kalama excepted, were upgraded to "increase log recovery, reduce [labor] cost, and produce the most profitable product mix." Pope & Talbot continued to be an industry leader in improving lumber manufacturing. The firm once again rebuilt and expanded Port Gamble. Its computerized headrig system returned 20 percent more lumber per log than its previous system. The rig scanned and positioned small logs at nearly exact saw tolerance. All the mills installed equipment to produce higher-value stress-rated (rather than visually graded) lumber for the prefabrication and component industries.

At a cost in excess of $10 million, the company rebuilt and enlarged the Oakridge veneer mill. A new steam tunnel conditioned logs for higher recovery value. Precision scanning and centering systems and a new lathe were added. An automated lumber sorting system costing $3 million was installed in 1980 in the big Oakridge sawmill, now capable of producing 140 million board feet and 80,000 bone-dry units (24,000 pounds each) of wood chips a year.

Also in that year Whelan established a $20 million base in ponderosa pine country. Pope & Talbot purchased and rebuilt a partially burned mill at Spearfish, in the Black Hills of South Dakota. Although the company had little experience in this part of the country, its knowledge of small-log technology counted heavily against less advanced local competitors. The tree types, geographic location, and low price also coincided with corporate goals. Warm, dry ponderosa, also called western yellow pine, grew at the extreme eastern limit of its natural range on nearby slopes in South Dakota and neighboring Wyoming. The trees were 100 to 160 feet in height and from 2 to 4 feet in diameter. Most trees grew, matured, and declined over a 125-year cycle. By 1980 the old-growth pine had largely given way to well-stocked and manageable second-growth.

Following multiple-use doctrine, Pope & Talbot applied at Spearfish "the beneficial use of our forestlands for recreation, wildlife protection and as watersheds" that had been celebrated in its 1974 annual report in recognition of changing attitudes toward forests. A commitment to sustained use prompted selective harvesting. Loggers removed mature trees and left the others to reseed and mature in turn. The Black Hills National Forest furnished 60 percent of the logs used at Spearfish, another company provided 28 percent, and small woodlots supplied the remainder.

Late in 1981 the company reopened Spearfish as a computerized, state-of-the-art sawmill able to withstand freezing winters. The improvements raised annual production from 70 to 108 million board feet, making Spearfish the largest ponderosa pine producer in the Black Hills. The mill turned out boards and high-quality lumber used for interior trim, cabinet work, and paneling. Sales focused on the home repair,

Kitchen cabinets display the use of Pope & Talbot hardboard made from mill and veneer waste at Oakridge in the 1960s.

remodeling, and do-it-yourself markets, all less volatile in prices than construction lumber. The plant served customers in the Midwest and on the East Coast.

The lumber industry could support the building of three million American homes a year by 1980, but this market demanded only half the existing capacity, even in good years. Pope & Talbot therefore redirected Oakridge to foreign business, which absorbed 60 percent of the output. Management also converted hemlock into specially cut lumber for Japan, where customers paid premium prices. Port Gamble, meanwhile, converted to both dimension-size and exportable metric-size lumber. By 1989 it too functioned largely as an exporter.

Committed to upgrading all product lines, Pope & Talbot added decorative interior panels to Kalama's textured exterior lines in 1976. The new Cedarstrip, promoted for its "natural" look, appealed to do-it-yourselfers and builders alike. The mill, though, lost much of its allure. Between 1970 and 1984 national plywood production slowed to a depressing rate, with demand expected to level off over the next ten years. High-speed, high-volume, and low-cost southern plywood mills made deep market inroads. Plywood output doubled in the South but dropped by a billion square feet on the West Coast. The Kalama plant also fell behind from a technical point of view, especially in pollution control equipment, and, like Port Gamble, it suffered from a lack of logs. Pope & Talbot therefore closed Kalama in 1979, abandoning the manufacture of plywood.

SPIN-OFF

LIKE MANY FOREST PRODUCTS CONCERNS, POPE & TALBOT UNDERWENT A MAJOR TRANSformation between 1940 and 1980. First, the company altered its business strategy from a resource orientation to a converting and marketing focus. Rather than boost sales of existing lines through lower production costs, it increased sales by developing and promoting a broader product mix in wood, pulp, and absorbent paper. Second, Pope & Talbot transformed its raw material base from an old-growth, large log emphasis to a focus on managed-growth small logs. Third, the firm shifted from technologies oriented to larger logs to those appropriate for smaller-sized timber.

Pope & Talbot by the early 1980s had survived while many competitors disappeared, but its future seemed uncertain. Management conceded that its market position and timber holdings were at low ebb. Outside advisers urged a spin-off of undervalued timber assets that tempted hostile takeover attempts. The biggest corporate buying and selling wave in American history was under way. A corporate raider's takeover, breakup, and closure of the Crown Zellerbach empire "really encouraged us to spin off," George Folquet recalled. Company executives thought it better to use the capital that could be generated from the undervalued timberlands by mortgaging them prior to the spin-off in order to diversify and modernize existing mills.

The separation of assets approved by stockholders in December 1985 was a major event in Pope & Talbot history. It transferred major properties to a limited partnership, Pope Resources, Inc., incorporated in Delaware. Peter T. Pope and cousin Emily T. Andrews were the principal partners. Stockholders received partnership units proportionate to their company shares and tax benefits. It was the first time an old-line wood products company gave shareholders essentially the choice of pursuing a trees-and-property development scheme or a manufacturing one. The Pacific Stock Exchange immediately listed Pope Resources as a separate concern from Pope & Talbot. The experienced Folquet became its president and chief executive officer.

Pope & Talbot transferred the firm's Washington timber—78,000 acres with an estimated 650 million

> *"After diversification the next step to increased public participation was to spin off our timberlands and property developments to shareholders, thus increasing their visibility. By putting the assets in a publicly traded partnership, Pope & Talbot would be less vulnerable to takeover bids, more tax efficient, and would give shareholders a more assured way of realizing the true value of their timber assets. The thinking was right: The two securities traded for more than the single security had. From this point on, the way was clear for Pope & Talbot to use equity financing to expand its business. Up to this point, Pope & Talbot was the only company on the New York Stock Exchange that had never had a public offering. It had financed itself entirely from internally generated funds. Until the spin-off only stock sold by family members had entered public hands."*
>
> —*Peter T. Pope*

board feet—and the Port Gamble mill site and town to Pope Resources. The new company also received Port Ludlow, then at a developmental standstill, and 4,400 acres of undeveloped land in Oregon and Washington. Altogether, Pope Resources obtained about 7 percent of the company's total worth, with a declared fair market value of $12.6 million. Pope & Talbot retained all manufacturing centers, logging rights in British Columbia and the Black Hills, and the 40,000 acres remaining in the Penn Tract. It leased Port Gamble to keep the mill running and undertook management of the town site.

The transaction substantially improved the balance sheet. Mortgaging the timber before the spin-off provided Pope & Talbot with $22.5 million to pay off high-interest loans. Long-term debt declined as a result, from 35 percent to 26 percent of total capitalization, one of the lowest rates in the industry. Company liquidity and shareholder equity were increased. Despite the wisdom of the move, Pope & Talbot managers had to accept that historical sources of responsibility, influence, and profit were gone. The company had redefined itself to function exclusively as a manufacturing entity. When rid of its last Penn Tract holdings, it would no longer raise trees, own forests, or engage in the land and resort development business. The nineteenth-century Walkers, Talbots, and Popes likely would have saluted Pope & Talbot's altering the course to survive. They would have pined for the long-lost ships, of course.

Peter T. Pope

CHAPTER EIGHT

NEW ERA

1986 AND BEYOND

OAKRIDGE

BY THE EARLY 1980S OAKRIDGE—THE MILL AND TOWN ALIKE—WERE IN TROUBLE. THE Pope & Talbot, Hines, and Bald Knob wood plants employed a thousand people in and around the community. Despite being too highly dependent on the wages and support services the mills provided, most Oakridge residents ignored warnings to broaden the local economic base. It was too late to do so when unemployment mounted to more than 15 percent after 1978, with a much higher unofficial rate. The town lost population. Changes in forestry practices, new environmental regulations, and altered markets had a tremendous impact on the companies that undergirded the community. Like the town, Pope & Talbot survived, but not as the operation it had once been.

In 1980, Pope & Talbot completed a major rebuilding and expansion of its Oakridge veneer mill and modernized its lumber mill sorting. Forty thousand acres of splendid old-growth forest in the Penn Tract supported the highly integrated operation. Digging its way out of overbid federal timber, the company blamed the swollen log prices for substantial and sustained losses at Oakridge. Meanwhile, growing environmental restrictions shrank its timber opportunities. Neither sagging lumber prices in the early 1980s nor expectations of a lower construction rate helped the situation. Late in 1980 the flagship mill closed temporarily. When it reopened, there were fewer weeks of full operation and, because of productivity gains, fewer employees were needed to maintain output.

Profitability was pushed largely by heavily cutting the company's own timber. Logging was accelerated on the Penn Tract to help finance the massive corporate expansion of the early 1980s. Between 1983 and 1987 the higher yield helped drive up overall corporate sales by 90 percent, earnings by 650 percent and pretax income by 800 percent. In 1987, Pope & Talbot was listed as a Fortune 500 company. Wall Street analysts rated the company bonds "investment grade."

But the company's prospects were gloomier than statistics and listings indicated. The Oakridge mill was on the verge of closure, its green veneer mill already leased to an independent operator. Management by the mid-1980s accurately anticipated sharp reductions in logging on the Willamette National Forest and other western lands. A piecemeal though fundamental shift was under way in resource stewardship.

The company tug *Sutherland* on Arrowhead Lake, B.C. The photo was taken by the tug's operator, Andy Hawkins.

Ecosystem management, not extraction, became the mission of public-land agencies. Environmentalist pressure had sharply mounted to create wilderness and roadless areas on public land and to preserve old-growth forests. Many Americans worried about protecting salmon and the northern spotted owl. The spotted owl issue was the final negative factor in the debate over Oakridge's future. "The spotted owl destroyed the chances to keep Oakridge going," Peter T. Pope explained, calling the highly publicized fight to save the species Oakridge's "death blow." Pope & Talbot expected that logging restrictions on state lands to protect the reclusive owls' habitat would not be far behind.

Labor relations of hard-hit North American lumber and pulp industries also grew more adversarial during this period. The Oakridge sawmill shut down during negotiations over company-requested reductions in wages and benefits. Oakridge had generated no profit since 1979, Pope & Talbot explained, and the future prospect was one of mounting losses. Its unions made large concessions to save jobs and protect the community. The sawmill reopened at full capacity in July 1986.

(left) Pope & Talbot executives gather at Oakridge in 1986 to watch the last old-growth tree cut from the Penn Tract. Pictured from left to right are Peter Pope, Jim Stout, Bob Madison, Dolph Andrews, Steve Mason, Cy Walker, Carlos Lamadrid, Stub Stewart, [unknown], Roger Hayes, Guy Pope, Harry Clark, Tom Faught, and Mike Flannery.

(below) Company advertising in Australia.

"This was a last gasp strategy," Pope recalled. "By cutting wages we kept Oakridge profitable for a short period of time." In line with the industry trend toward higher-value products, the reopened mill replaced most lower grades with high-priced lumber. Like many private-forest owners, Oakridge and Port Gamble also attempted to export their way out of weak domestic markets, mainly by sending old-growth lumber and logs to Asia. Shipments abroad amounted to 60 percent of Oakridge's sales by 1988, enabling it to contend with considerable domestic problems and to lead all company mills in sales. Still, the mill itself was not considered economically viable. Production costs and wood costs were simply too high. Pope & Talbot felled the last of its own nearby mature timber in 1987. Without the cheap Penn Tract logs, mill production fell by 28 percent the next year.

In the 1980s Pope & Talbot management felt company forest plans and practices endured harsher public scrutiny and government environmental oversight than the company experienced in the 1970s. After three years of restrictions on public timber sales in Oregon and Washington and unwanted increases in the cost of the remaining timber, management in 1989 formally shifted the corporate lumber emphasis to the Black Hills of South Dakota and to British Columbia, "where there is assurance of raw materials." It scheduled the last 31,406 acres of the Penn Tract for sale, netting the company $24.8 million by 1993. It took a jaundiced look at a Port Gamble mill crippled in its ability to obtain adequate timber. And in 1989 management sold the Oakridge

OAKRIDGE **147**

sawmill and veneer complex to the Bald Knob Lumber and Timber Company, for an after-tax gain of $3.5 million. Bald Knob reopened a modified plant, with 140 of the 500 employees once working there and in the woods. Bankrupted two years later, the unfortunate new owner closed the mill for good.

PORT GAMBLE

OAKRIDGE WAS NOT THE ONLY COMPANY MILL FACING TOUGH TIMES. HIGH LOG COSTS and reduced timber supplies slowly choked Port Gamble. Pope & Talbot spent $3 million to $4 million a year on its modernization during the mid- and late 1980s. Capacity was increased from 86 million to 150 million board feet. An efficient workforce using modern lasers, computers, and cutting equipment focused on 4- to 30-inch diameter logs. In 1991, Port Gamble employees perfected the country's first industrial-sized wood pellets. As a means of improving air quality, the federal government encouraged the use of the pellets in pulp mill boilers.

Nevertheless, the fountainhead mill fell victim to the same elements that would lead to the closure of 167 other Oregon and Washington mills between 1990 and 1995. Following the Oakridge sale, Pope & Talbot wrote down the Port Gamble facility on its books to salvage value—a $11.9 million net loss. After a monthlong shutdown the mill reopened with only one shift, and even that was contingent on log availability and cost.

Government assistance was needed, the local Port Gamble manager argued, to counter the impact of regional log exports. As a counter to the high-priced Japanese market, the former exporter joined other Washington operators in securing a ban on the exportation of logs from state-owned land. However welcome, this stopgap action failed to save the mill. Even if the plant were able to purchase all nearby harvestable timber, it would have been able to fill no more than 15 percent of its orders. Availability of state and federal logs on the Olympic Peninsula, meanwhile, declined sharply. In a seller's market timber costs escalated during the mid-1990s. Port Gamble produced only 87 million board feet in 1994 and a disappointing 53 million board feet in 1995.

In one more sign of a passing era, Pope & Talbot announced the closure of the oldest continuously operating sawmill in North America in the summer of 1995. "Except for a ten-month period, the mill has operated at a loss in the 1990s," management reported. Port Gamble and Oakridge had "survived because we had some timber for them," Pope recalled. "When that began to run out, and when the public timber wasn't available, they went down." The industry, he added, was "a raw-material-driven survival business" and, in fact, a historically "transient" enterprise.

In better times more than two hundred people had worked in the mill. Only ninety-six were on hand at the time of the closure. Thirty-five families continued paying rent for simple homes heated by woodstoves. Many of those laid off at Port Gamble came from generations of local mill workers. Until recently, employment had generally been steady. Their days had been marked by mill whistles, the rumble of logs, and the whine of saws. Laborers smelled fresh-cut wood and inhaled sawdust, moved along trembling walkways, and felt the sparks of grinders. All that remained was Historic Port Gamble, the living Pope Resources museum, and a number of residences.

The shut-down Port Gamble mill, seen from Little Boston, awaits site clearance in August 1996 after more than a century of continuous operation. *(Tony Johnson, photographer)*

CONSUMER PRODUCTS

POPE & TALBOT NOW PROMOTED ITSELF AS "A CYCLICAL, SOLID-WOOD PRODUCTS COMPANY [that] has been transformed into a balanced forest products company heavily weighted toward consumer products." Americans were predicted to spend in 1990, according to contemporary studies, $8.5 billion on paper tissues and diapers. Reflecting the importance of tissue sales, William A. Whelan had been replaced as president and chief operating officer in 1984 by R. Steven Mason, an experienced consumer products executive. When Mason left the firm in 1990, Peter T. Pope again became corporate president.

The company expanded its tissue holdings in 1988, paying $24.5 million for a 48,000-ton mill in Ransom, Pennsylvania. The facility converted old milk cartons, cups, and other material into fiber for paper towels, napkins, and bathroom and facial tissue. As in the two Wisconsin mills, the machinery was small and slow but suitable for making short-run batches of store brands. The three mills together gave the corporation the largest private label and generic tissue capacity in the United States. East of the Rockies, Pope & Talbot held a 33 percent share of the market in 1990 after expanding the Ransom plant. But as Peter Pope noted, there were "not a lot of acquisitions opportunities left." In particular, the lack of a western mill prevented company exploitation of a full-fledged nationwide tissue market.

In contrast, disposable diapers had national potential at a time when plant acquisition costs were low and many brands of diapers had been driven from retailer shelves. Pope & Talbot entered this market in a big way in 1988, immediately ranking as the nation's second-largest private-label disposable diaper supplier. Filled with fluff pulp, baby diapers were easy to make but costly to ship. The firm therefore opened facilities across the country. In addition to opening a plant in Shenandoah, Georgia, the company bought four Georgia-Pacific plants: in Aiken, South Carolina; Maryville, Missouri; Oneonta, New York; and Porterville, California. Together, the mills supplied Albertsons, Target, Wal-Mart, and other major retailers.

Although the inefficient South Carolina plant was sold in 1989, the Consumer Products division remained the company's largest, generating almost half the corporate revenue. Competitors responded aggressively. In 1988, Kimberly-Clark launched a price and promotion war. Fort Howard continued as a fierce price-cutter. Falling tissue prices and reduced retailer interest in store-brand diapers raised additional obstacles to continued profit. The supermarket chains and the new warehouse stores squeezed better terms from their suppliers. Pope & Talbot was forced, according to a 1988 employee newsletter, to provide these retailers costly "marketing, research, product and packaging development, computerized space management, and profit planning as well as sales support and counseling." It partially duplicated these services for pulp and lumber buyers facing their own problems.

Pope & Talbot, so adept at prospering in slower changing and nonconsumer commodity markets, had no choice but to adapt to the fast-paced changes in the world of consumer products. The plants adopted the latest technologies for processing, tracking, and delivering orders. To quickly update inventories, data systems were linked to retail store scanners. Retailers desiring "just-in-time delivery" stockpiled smaller amounts of product and demanded quicker service. Lowering its own expenses, Pope & Talbot improved mill efficiency and cost controls and reduced inventories.

Rapid product innovation became a corporate routine. Field employees had once been assured that the diaper was "pretty well perfected" in its present form, hence "further change [was] unlikely." Instead, competition forced annual modifications that made the previous year's product seem like ancient history. Pope & Talbot replaced bulky diapers with very thin ones. It added moisture barriers, Velcro tabs, and elastic waist and leg bands; it introduced boy/girl and age-specific diapers and training pants.

In 1992, Spearfish began transforming the wood by-products that had previously been burned into 20,000 tons a year of environmentally friendly wood pellets, used by woodstove owners.

Some of these changes created unanticipated challenges for the firm. The new "ultra thin" diapers used super-absorbent polymers, a component that reduced demand for pulp and compromised the corporate goal of economically integrating the pulp and diaper mills. Development of the adult disposable diaper Ensure, however, opened a new market sector. Frequent changes had to be made in the structure, design, and packaging of products, too. Equipment had to be retrofitted. By the early 1990s entirely new facilities and upgrades were needed to counter industrywide cost cutting. Ransom got a new converting and distribution center, and Eau Claire a $22 million pulp mill modernization.

In a harshly recessionary 1991, a management prideful of the company's conservative financial history reported a financial hemorrhage. The losses were attributed to several causes: absorbent paper overcapacity and price wars, acquisition costs, an upward spike in chip prices, and the cost of major pollution abatement projects. The company had weathered severe crises before, but now had no rich timber to fall back on.

A major corporate restructuring was undertaken. The number of salaried employees was reduced by 20 percent in 1992; the hourly workforce by 11 percent. Reforms in management style complemented the downsizing. Like much of corporate America during the 1980s, Pope & Talbot introduced new working practices, simplified managerial structures, and reduced the layers between its own raw materials and outlets. An annual report spoke of the firm attempting to move "from being control oriented to being involvement oriented." Executives were encouraged to act like coaches, and employees were encouraged to think for themselves.

Consumer product profits, however, failed to rebound. Between 1992 and 1994 tissue losses mounted. Procter & Gamble, in particular, intensified the ongoing price war. Pope & Talbot continued to cut costs to meet the challenge. At Ransom in 1995 efforts to curb seeming labor inefficiencies provoked a seven-month strike. The plant maintained limited production, relying in large part on salaried employees. Strikers, worried that the plant might close permanently, returned to the job on terms favorable to the firm. A hundred fewer workers were needed, a reduction that restored the mill's profitability.

Losses of nearly $25 million in 1995 were the highest in corporate history. The lumber and absorbent product divisions suffered the most. High world pulp prices and low raw material costs enabled the pulp and paper division to avoid major losses. Wall Street rating agencies reduced corporate bonds to below-investment grade. In 1996 the banks tightened long-term borrowing requirements in a way that weakened the stock's main attraction: Pope & Talbot had paid hefty dividends for a quarter century. In the 1970s it had established dividends as a percentage of beginning-of-the-year equity, rather than a percentage of earnings. The directors intended to ensure steady dividends, despite cyclical markets.

Everybody must "adjust" to fit the "reality" of record losses, Michael Flannery told employees in May 1996. After a decade as a company wood product executive, Flannery, a Harvard M.B.A., had replaced Peter T. Pope the previous September as president and chief operating officer. Pope remained as chairperson and chief executive officer. Both stressed the need to rebuild the corporation.

In the midst of a trend toward manufacturer consolidations, Pope & Talbot quit the diaper field. Because thinner diapers had reduced freight costs, a few large facilities made greater economic sense than a large number of scattered small plants. A big new investment to compete more efficiently was deemed too risky. Early in 1996, Pope & Talbot sold its inefficient diaper plants to Paragon Trade Brands for $65 million.

Tissue industry overcapacity stabilized in 1996 and 1997, earning profits for the first time since 1991. Nevertheless, Pope & Talbot had decided with the diaper plant sale to make enough minor changes to the tissue plants to attract a buyer and then exit this niche market. Otherwise, management concluded, costly investments in marketing and product innovation would be required to remain competitive. The sale in 1998 of the Pennsylvania and Wisconsin mills to an investor group for $147 million provided funds for new investments. "Because we were a small tissue producer in the land of giants," Pope told the Portland *Oregonian* in June 1999, "we weren't able to achieve [profit] margins that were adequate." With the sale of the tissue and diaper operations, Pope & Talbot once again became exclusively a lumber and pulp manufacturer. "The company is smaller now," Flannery announced. "It's almost a new company."

TREES AND MILLS

COMPANY PURCHASE OF A SAWMILL NEAR THE SOUTHERN END OF THE BLACK HILLS IN 1989 locked in more long-term timber supplies within a cutting circle shared with Spearfish, sixty miles away in South Dakota. Costing $2.7 million before upgrading, the small Newcastle, Wyoming, operation handled ponderosa pine according to traditional sustained-yield doctrine. By 1991 the two mills manufactured a quarter of all company wood products. The high-value Midwestern home repair and remodeling markets were their major outlets.

In 1992 the company increased Canadian operations, becoming the biggest firm in the Boundary Timber Area and the largest employer in three provincial towns within driving distance of one another. The firm's tested strategy of acquiring assets inexpensively at the end of the business cycle was again employed. Nobody but Pope & Talbot wanted the Castlegar mill, with its 225 million board feet capacity, in a time of depressed lumber markets. The $19 million transaction paid off handsomely, however. "We more than earned the purchase for the facility during the first twelve months of operation," the 1993 annual report noted.

Castlegar increased British Columbia production to 435 million board feet, and total corporate mill capacity by 40 percent. The province provided 85 percent of the firm's wood products revenue from 1992 through 1994. The company invested $7.5 million in expanding and retooling the mill. Sophisticated new machinery made both one-inch and two-inch boards and shifted back and forth from 24-foot lengths to 20-foot-long, machine-stressed lumber.

Once Port Gamble closed, the Boundary mills provided about 75 percent of total sawmill capacity.

In the Black Hills's short-log country trees are often small enough for one person to run a single machine that operates as faller-buncher, bucker, delimber, and stacker. Only mill transport and cleanup are then required.

Employing eight hundred people, including seventy-five loggers, they were the largest source of jobs in their immediate areas. Castlegar ownership added 1.4 million acres in long-term logging rights to existing company resources. About 3 million acres of license holdings permitted almost continual rotation of cutting among mature stands in the vast Boundary Timber Area.

British Columbia expected tenure-holders to act as caretakers of the land. The Forest Act of 1978, an offshoot of the provincial "green revolution," mandated long-term resource plans, detailing strategies for harvesting, reforestation, and silviculture. Company crews instituted a three-year creek restoration project, the first in interior British Columbia. Other crews planted millions of seedlings a year and tended young and maturing forests. Additional environmental requirements increased operating costs. A new Forest Practices Code phased in stricter regulations between 1996 and 1998. The "annual allowable cut" was reduced, lowering harvest volume, and reforestation requirements were tightened.

Corporate leaders, though hardly enthralled by the new regulatory climate north of the border, maintained faith in long-term Canadian resource stability. Restrictions still seemed a good deal less onerous than in American jurisdictions. Canada, Pope assured the stockholders in April 1997, was "a forest industry

Midway, pictured, and the Grand Forks mill are two of Canada's more efficient sawmills. Midway is among the world's largest producers of machine stressed–rated lumber, evaluated by equipment to guarantee its working stress, elasticity, and visual grades.

> "One of my great disappointments as CEO of Pope & Talbot was not stabilizing the earnings of the company. When we purchased the Halsey, Oregon, pulp mill, its tissue-grade pulp was difficult to sell. In what we saw as a solution and an opportunity, we entered the private-label (store-brand) tissue and diaper business by buying the absorbent products division of the Brown Paper Company. This, we hoped, would add stable consumer products to our volatile commodity products. At that time private-label products were in great demand, and we achieved record profits. Brand-name companies, however, counterattacked by slashing their prices, which forced us to do the same. Next, the lower-cost Fort Howard Paper Company became a private-label competitor. We struggled to fix this business for more than ten years but finally gave up. We got our investment back when we sold our absorbent products division, but it cost us ten years—ten years in which we might have been growing into something with a real future."
>
> —*Peter T. Pope*

Under computer control one of Grand Forks's two optimizing board edgers scan, position, and edge two thick slabs.

friendly environment." By contrast, the Department of Agriculture, parent of the U.S. Forest Service, "didn't care about the Pacific Northwest because there's more votes in the East," he stated in an interview with a B.C. newspaper. "And [back] out East they'd rather have trees out here than jobs."

Neither the British Columbia nor the Wyoming or South Dakota mills served offshore markets. This narrowing of focus marked another significant historical change. The earlier Pope & Talbot sawmills had always sought at least secondary markets abroad. For more than a century the company had sent long-voyagers to the Pacific, into Asia, to eastern South America, around Cape Horn, and through the Panama Canal to the Caribbean, the eastern United States, and to and from Europe.

In wood products the firm was now primarily a Canadian exporter and secondarily, in terms of volume, a domestic operator in the Black Hills. (The exception was wood chips, which went to a Japanese pulp maker on long-term contract.) As

Workers along the "dry chain" conveyor in the Grand Forks's planer building handsort low-volume planed lumber. A computerized sorter, beside the upper catwalk, automatically sorts high-volume products.

a multinational corporation, Pope & Talbot necessarily attended closely to U.S.-Canadian relations; to currency, regulatory, and tax differences; and to trade disputes. One major commercial conflict hit home: the issue of whether Canadian lumber competed unfairly with American lumber. In particular, firms in the southern United States demanded restrictions on lumber imports from British Columbia. The southerners argued that provincial forestry policies amounted to a subsidization of prices. Canada capitulated and offered a concession in 1986, imposing a 15 percent lumber export tariff. Five years later, in 1991, the U.S. Congress actually imposed a 6.51 percent duty, affecting the 500 million board feet Pope & Talbot sent across the border that year. The tariff, happily for the firm, was rescinded in 1994. Under a 1996 agreement between the two governments, duties were reinstituted on shipments in excess of company-by-company quotas. In subsequent years Pope & Talbot temporarily closed its B.C. mills to stay within the agreed-on duty-free limits.

PULP

SELLING PULP TO THE ADJACENT JAMES RIVER PAPERMILL AND WET LAP PULP TO Northwest newsprint makers, the Halsey mill "has become probably one of the most profitable pulp mills in the West," Pope told industry analysts in May 1990—just before international prices nosedived. Management faced a greater market gyration than any yet experienced. Global pulp earnings were, at best, flat throughout the 1990s.

At Halsey, production occasionally fell below capacity. The near-collapse of some Asian economies reduced American pulp and paper exports. And domestic customers shifted to alternate materials. Concerned

The pulp-making sequence.

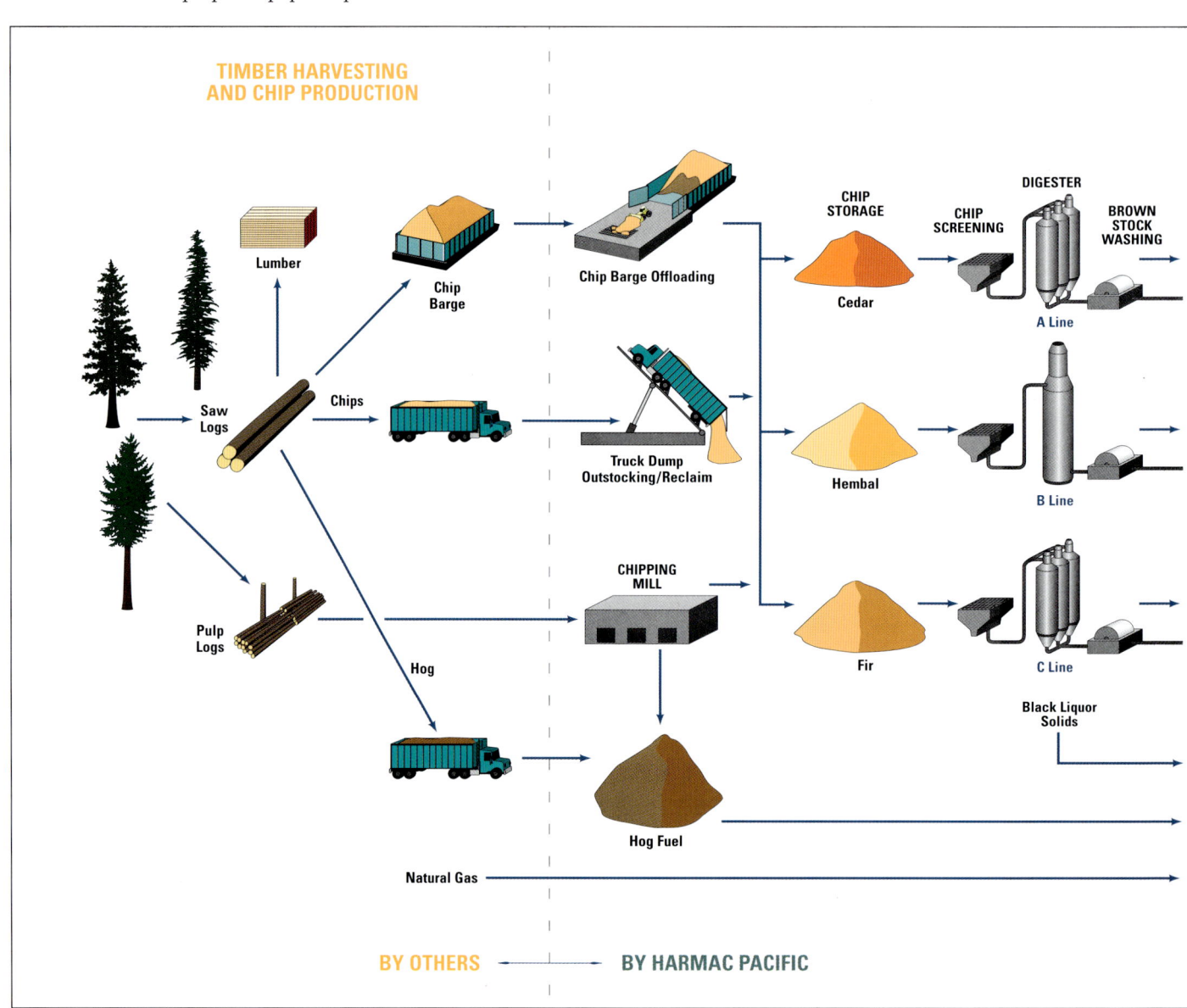

about the environmental effects of gorged landfills, the adjacent James River papermill switched from Pope & Talbot pulp to wastepaper pulp, and newsprint makers turned from the firm's virgin pulp to pulp from recycled newspapers. Between 1992 and 1998 total pulp losses mounted to $37 million. In a glutted world market the ten-year cycles of pulp and lumber now counterbalanced one another only 80 percent of the time, running against the countercyclical corporate approach two out of ten years.

Huge investments to transform Halsey and to meet stricter environmental regulations there pushed down profit margins. Pope & Talbot spent $70 million upgrading the facility between 1989 and 1994. The company almost completely reconfigured Halsey as the recession shrank markets for wet lap pulp, or White Gold 49. The mill shifted to production of higher-quality pulp for worldwide sale. Fully participating in a

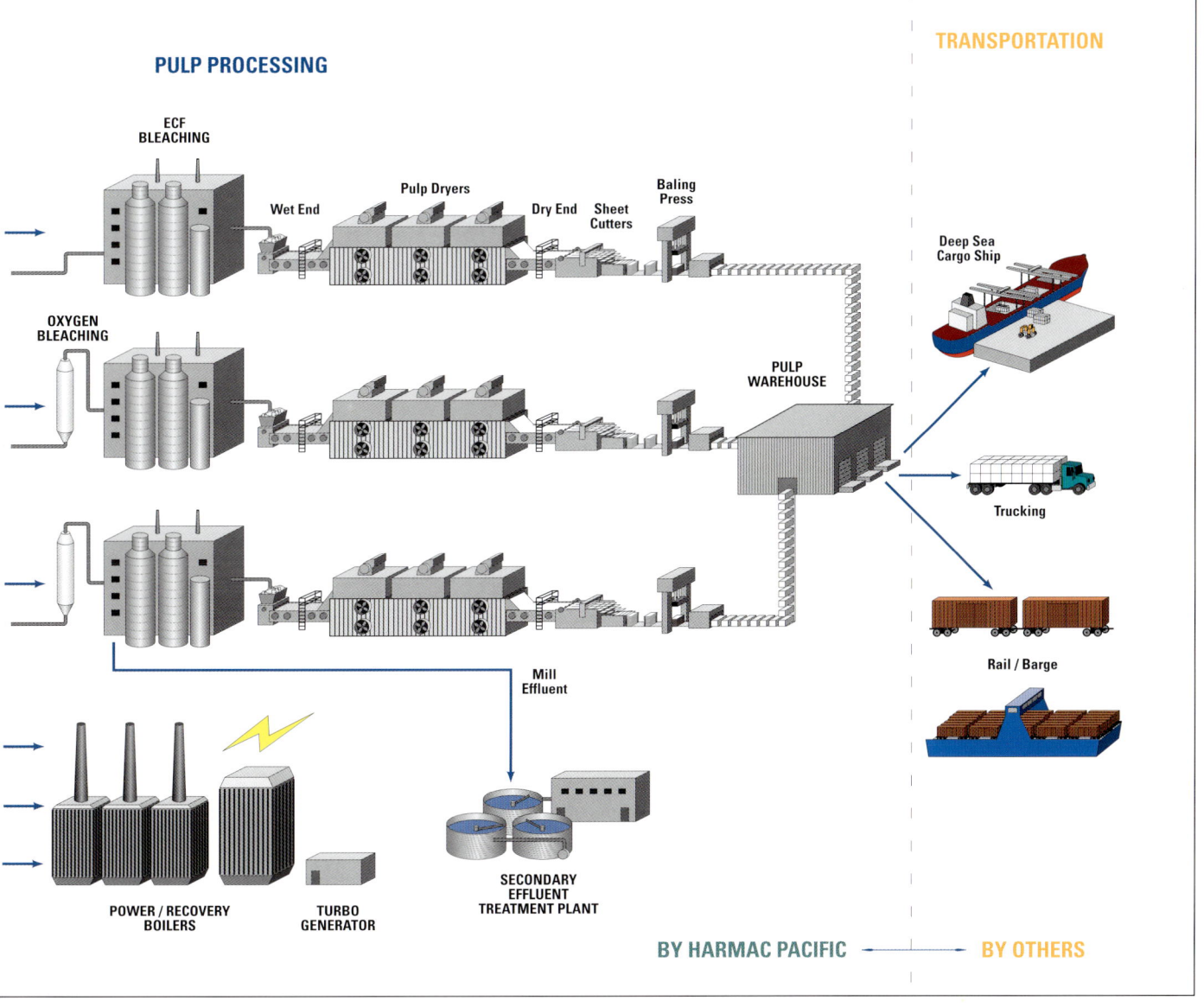

MODERN LUMBER MAKING

THIS CUTAWAY ILLUSTRATION (OPPOSITE PAGE) reveals midway's major manufacturing steps prior to lumber sorting. Front-loaders carry logs to three drive-on log decks. A crane straightens some logs. From the two small log decks (far left), logs drop down to one of three parallel lines moving into the plant. The third drive-on deck handles only large logs, sending them through a conventional headrig, which uses the largest gang saw.

Sawmills face the eternal problem of sawing a maximum amount of rectangular-shaped boards from an essentially round log. Large circular cut-off saws on three parallel lines *(in front of the control booth)* cut ("buck") the logs to specific lengths. Crooked and eccentric logs fall into a rejection bin. Debarkers strip off bark, used as hog fuel to fire furnaces or boilers. Other sawmills debark before bucking, or cut boards to length only after drying. No two modern sawmills are exactly alike.

Acceptable logs continue through a primary breakdown machine. Here, they pass slabber heads, which chip wood away to create a flat outside surface. They next pass through band saws, which cut one- to two-inch-thick slabs off each log side. These "sideboards" go through an optimized board edger, which squares the edges and creates rough-cut boards. The remaining "center cant" goes downstream to a gang edger. In one pass its multiple saws cut the cant into many pieces of rough lumber.

After the board or gang edger, lumber flows to one of two identical sets of floor chains. Employees manually inspect lumber on each line. They divert marginal boards to reprocessing by the board edger. Good lumber pieces merge and continue through an unscrambler, which places the lumber, one piece at a time, onto a lugged chain. Each piece is again scanned, computer optimized (to maximize the yield), and automatically trimmed to length before flowing to the sorter.

Within the sorting building *(covered-over top right)*, each board is automatically sorted for thickness, width, or length, and flows into one of eighty bins. As each bin fills, its lumber is sent to a semiautomatic stacker, where small strips of lath are inserted between layers so air can flow in during the drying process. A forklift transfers the stacked green lumber to Midway's adjacent drying kilns *(not pictured)*. Environmentally controlled kilns dry the stacked lumber for two or three days. A forklift transfers each dried stack to the planer building *(not pictured)*. High-speed knives smooth the rough board surfaces into finished-product quality. Individuals grade each board. The boards are automatically retrimmed, re-sorted, and packaged for railcar shipment.

Lines running somewhat parallel to the sorting building send by-products—chips, sawdust, and hog fuel—to elevated bins for truck loading near the log yard. Waste burning in the outside, high wigwam has now ended.

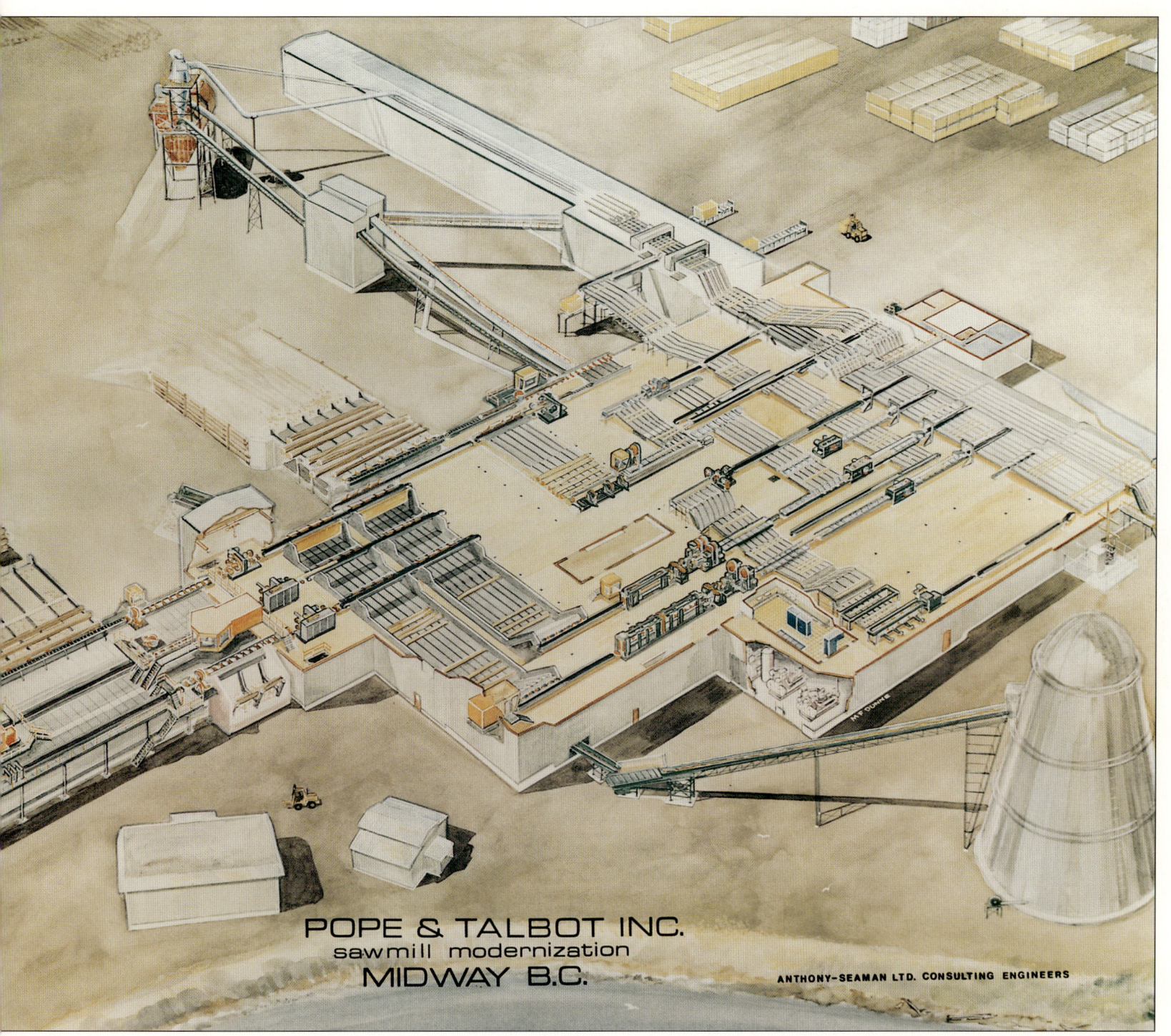

Tons of baled pulp load into vessels at Harmac Pacific's tidewater dock in British Columbia. Ships, boxcars, barges, and a new rail ferry afford the vast mill access to the world's fiber markets.

global marketplace, Pope & Talbot confronted bare-knuckled competition from mills in Brazil, the southern United States, and elsewhere. Halsey had difficulty realizing profits because it now made a Douglas fir pulp that was less valuable than northern bleached softwood kraft pulp (NBSK) made by competitors who enjoyed the economies of scale of their giant pulp mills. More than ever, corporate earnings were hostage to changes in world pulp inventories and demand.

Late in 1999, the company sold Halsey to a group of financial institutions in a complicated $64.6 million sale, lease, and buyback transaction. Pope & Talbot agreed to lease and operate the pulp mill and to legally reacquire it in either 2007 or 2012. Complicated financial advantages, including otherwise unobtainable pollution control tax benefits, generated this unique arrangement. When fully implemented, the technical upgrade would make Halsey compliant with all Environmental Protection Agency (EPA) rules. The transaction helped furnish $52 million toward buying majority ownership in Harmac Pacific, a giant pulp mill on Vancouver Island, British Columbia.

Halsey's transfer completed the final step in the repositioning of Pope & Talbot as an exclusively wood product and pulp business. Twenty years after first entering the pulp industry, the company took advantage of a strong downturn in prices to gain control and then total ownership of Harmac Pacific, whose single plant made it one of Canada's largest pulp producers. In January 1998, Pope & Talbot paid $53.4 million for a majority share of Harmac's waterfront mill at Nanaimo, British Columbia. The next year it completed the acquisition for $20.4 million. The mill's three lines turned wood chips into high-grade NBSK pulp. NBSK fiber is stronger, longer, and thinner than other varieties of pulp and is in increasing use worldwide. Harmac Pacific pulp sold well in the United States and Europe, in such niche sectors as the market for lightweight magazine paper.

This specialty mill added to the company base in British Columbia. The acquisition tripled total pulp capacity and raised the profile of pulp within the corporation. The Halsey and Nanaimo mills furnished 49 percent of corporate revenues in 1999, compared with 25 percent two years earlier. Late in 1998 the Harmac Pacific administrative, sales, and marketing offices were consolidated in Portland. By 1999, Pope & Talbot was the world's ninth-largest producer, at 445,000 metric tons, of NBSK.

EPILOGUE

"MY ROLE IN THE COMPANY REALLY [HAD BEEN] TO TAKE US FROM A FAMILY COMPANY to a public company," Peter T. Pope explained, and to being a "more professionally run company" while holding the firm together.

In 2001 company president and CEO Michael Flannery also became chairperson of the board. Maria M. Pope, Peter Pope's daughter, remained the only member of the original families actively involved in the top leadership echelon (Peter Pope had retired in August 1999). Equipped with a Stanford M.B.A., she rose from a planning and budgeting manager in 1995 to secretary-treasurer and in 1999 to vice president and chief financial officer. "There's no way you can separate" the Popes and the company, she said.

The 150-year-old concern carefully retained its corporate independence, maintained its exclusive focus on lumber and pulp, and demonstrated its distaste for the mergers and large-scale acquisition deals that attracted its competitors. Managers focused on survival, but not by hiding from challenge. They were prepared to seize opportunities and take advantage of still highly volatile conditions. Serious economic difficulties abroad, the onset of a U.S. recession, rising energy costs, and restoration of heavy U.S. import duties on Canadian lumber took their tolls. Literally. For Pope & Talbot, which shipped about three-quarters of its

> *"Today Pope & Talbot is a truly public company, with a board made up primarily of outside directors and active trading in its stock. Management is selected by ability, not by bloodline. The company is small in a land of giants, but market share is not relevant because of the international commodities the company manufactures. Pope & Talbot mills are just as cost-effective as those of our larger competitors. After more than a century and a half, Pope & Talbot is the oldest wood products company in America. What I did was to take it from a family company to a public company and guide it toward becoming a more professionally run company. I am proudest that we have survived, and that I have been an instrument in the company's survival."*
>
> —*Peter T. Pope*

The current Pope & Talbot management team. Pictured from left to right are Abe Friesen, Maria Pope, Mike Flannery, and Angel Diez.

annual 600 million board foot lumber output to the American market, there was an onerous $10.2 million import duty in 2001 and significant sawmill downtime in British Columbia to avoid higher duties.

Profits of $32.6 million in 2000 turned into losses of $24.9 million a year later. Pulp revenue shot up in 2000, thanks to repeated increases in world pulp prices, low industry inventories, and the added Harmac Pacific production. When pulp prices and demand rapidly tumbled in 2001, Pope & Talbot temporarily reduced output at the 200,000 metric ton Halsey mill—and bought another pulp mill. Once more, the corporate leadership purchased significant potential at a bargain price during a declining market. The Mackenzie mill in northern British Columbia, which opened in 1972 and was modernized in 1996, cost Pope & Talbot only $104 million. Mackenzie added 230,000 metric tons of NBSK pulp capacity, giving the firm an overall 780,000-ton pulp capacity.

Management basically adhered to tested policies in 2000 and 2001. If fitting assets became available at a cycle's low point, acquire them. Finance the company conservatively, as in using $36 million from the final stage of the Halsey sale-leaseback to reduce corporate debt. Keep the mills at operational peaks while prudently assessing their resource use, raw material base, and market profitability. And improve the natural environment in ways that build corporate prosperity and customer and public trust.

In January 1998, Pope & Talbot paid $53.4 million, including debt, for a majority interest in Harmac Pacific. The next year it completed the acquisition for $20.4 million, including debt, by buying out the minority owners. But efficient management could not solve timber supply problems, which continued to plague U.S. operations. When the government stopped the sale of logs from the Black Hills National Forest in late 1999, it foretold the choking off of an essential supply to the sawmill in Newcastle, Wyoming. Pope & Talbot closed the 30 million board foot sawmill in 2000, partly offsetting this lost production through its recently upgraded 120 million board foot mill in Spearfish, South Dakota.

Smarter use of raw materials, less waste, and cleaner-running and more energy-efficient equipment continued to realize bottom-line social and corporate benefits. In 2000, Halsey's new chlorine dioxide bleaching facility exceeded federal air and water standards. Both it and the Harmac mills were 100 percent elemental chlorine–free operations. Having ultimately cost $38.6 million, the Halsey improvement forecast annual savings of $7 million in operating expenses. At Spearfish new control equipment reduced air particulate emissions to far below federal and state requirements. Waste bark fired the drying kiln at

Castlegar, British Columbia, ending expensive natural gas usage. Log recovery figures were impressive. Company sawmills discarded less than one-half of 1 percent of fiber from a log.

Pope & Talbot exceeded the newest Forest Practice Code requirements on its more than 3 million acres of British Columbia leases, where 51 percent of the tracts qualified as park, nonforest, and environmentally sensitive land. In 2000 independent auditors certified that Pope & Talbot's management plans in the province met the International Standards Organization's criteria for "green" planning, road construction, timber harvesting, transportation, and reforestation. Forest crews planted 6 million seedlings that year alone. Afterward, Pope & Talbot strove for sustainability and market approval within ever more rigorous international environmental standards.

Pope & Talbot may have long since abandoned its shipping business, but the company was running a tight ship, and sailing smoothly as it cruised toward its hundred and fifty-third year.

DESCENDANTS OF DEACON PETER TALBOT AND COLONEL WILLIAM POPE INVOLVED WITH THE MANAGEMENT OF POPE & TALBOT & POPE RESOURCES

GENERATION NO. 1

DEACON PETER TALBOT [Born March 29, 1783, East Machias, Maine; died July 21, 1875, Providence, Rhode Island] He married ELIZA CHALONER September 28, 1813 in East Machias, Maine, daughter of WILLIAM CHALONER and MARY DILLAWAY. She was born August 29, 1785 in Machias, Maine, and died May 22, 1831 in East Machias, Maine.

- **CAPTAIN WILLIAM CHALONER TALBOT** [Born February 28, 1816, Machias, Maine; died August 06, 1881, Astoria, Oregon] He became Andrew Pope's partner when his brother Frederick decided to return to the East. He sailed to San Francisco with a load of Maine lumber around Cape Horn.
- **FREDERIC TALBOT** [Born February 26, 1819, East Machias, Maine; died December 20, 1907, Providence, Rhode Island] He and Andrew Pope came to San Francisco by ship via Panama.
- **EMILY FOSTER TALBOT** [Born October 03, 1820, East Machias, Maine; died February 1906, San Francisco, California] She married Andrew Pope, further strengthing the family bonds.

COLONEL WILLIAM POPE [Born March 30, 1787, Charlston, South Carolina; died November 06, 1864, Boston, Massachusetts] He married PEGGY DAWES BILLINGS September 27, 1810 in Boston, Massachusetts, daughter of WILLIAM BILLINGS and LUCY SWAN. She was born March 06, 1788 in Boston, Massachusetts, and died September 08, 1862 in Boston, Massachusetts.

- **ANDREW JACKSON POPE** [Born January 06, 1820, East Machias, Maine; died December 18, 1878, San Francisco, California] He helped found Pope & Talbot with financial backing from William Pope & sons, his father and brothers.

GENERATION NO. 2

ANDREW JACKSON POPE married EMILY FOSTER TALBOT September 06, 1852 in East Machias, Maine, daughter of PETER TALBOT and ELIZA CHALONER. She was born October 03, 1820 in East Machias, Maine, and died February 1906 in San Francisco, California.

- **GEORGE ANDREW POPE SR.** [Born 1864, San Francisco; died October 16, 1942, San Francisco] He was an investor and society leader in San Francisco and ran Pope & Talbot after his cousins W. H. Talbot and F. C. Talbot retired.

CAPTAIN WILLIAM CHALONER TALBOT married SOPHIA GLEASON FOSTER May 26, 1846, in Eastport, Maine, daughter of EZEKIEL FOSTER and RUTH HAYDEN. She was born October 15, 1823 in Eastport, Maine, and died March 18, 1911, in San Francisco, California.

MARY ELIZABETH TALBOT [Born March 22, 1847, Eastport, Maine; died 1894, San Francisco, California] She married Henry Dutton Jr., who worked for Pope & Talbot.

EMILY FOSTER TALBOT [Born September 12, 1848, East Machias, Maine; died February 20, 1922, San Mateo, California] She married Cyrus Walker, manager of the Northwest Mills.

SOPHIA GLEASON TALBOT [Born June 01, 1850, Eastport, Maine; died 1929, San Francisco, California] She was the grandmother of Harriett Brownell, who married George Pope Jr.

WILLIAM HAYDEN TALBOT [Born May 24, 1858, San Francisco, California; died 1930] He was the leader of Pope & Talbot for many years following the death of the two founders.

FREDERIC CHARLES TALBOT [Born May 18, 1860, San Francisco, California; died 1919] He helped his brother run Pope & Talbot.

FREDERIC TALBOT married HANNAH SANBORN 1846, daughter of CYRUS SANBORN and SUSAN GARDNER. She was born January 26, 1825, in East Machias, Maine, and died January 26, 1854, in New York.

CHARLES FRANCIS ADAMS TALBOT [Born August 22, 1848, East Machias, Maine; died July 14, 1931, San Francisco, California] Even though his father returned to the East, he came west and helped his young cousins run the company. He was an executor of both Andrew Pope's and Capt. W. C. Talbot's wills.

GENERATION NO. 3

EMILY FOSTER TALBOT married CYRUS WALKER April 30, 1885 in San Francisco, California, son of JAMES WALKER and ELIZA HEALD. He was born October 06, 1827, in Madison, Maine, and died October 01, 1913, in San Mateo, California. He purchased 10 percent of Pope & Talbot and was instrumental in the success of the Northwest Mills.

TALBOT CYRUS WALKER [Born 1886, San Francisco, California; died 1956, Santa Barbara, California] He retired in reverse, starting to work at an older age when he was needed during World War II.

SOPHIA GLEASON TALBOT married IRA PIERCE July 27, 1876 in San Francisco, California, son of WILLIAM PIERCE and MEHITABLE CHARLES. He was born December 26, 1836, in Standish, Maine, and died March 13, 1913, in San Francisco, California.

SOPHIA GLEASON PIERCE [Born August 04, 1879, San Francisco, California; died September 28, 1975, San Francisco, California] She was the mother of Harriet Talbot Brownell, who married George A. Pope Jr.

WILLIAM HAYDEN TALBOT married ANNIE DOUGLAS LAMONT in 1887. She was born 1862 in San Francisco, California, and died in 1915.

FREDERICK C. TALBOT [Born November 1895, San Francisco, California; died 1955, San Francisco, California] He worked many years for Pope & Talbot in the Northwest as well as in San Francisco.

CHARLES FRANCIS ADAMS TALBOT married AMY NORTON BOWEN November 22, 1877, in Oakland, California, daughter of PARDON BOWEN and FRANCES BLANDING. She was born February 18, 1857, in Stockton, California, and died October 02, 1944, in Washington, D.C.

AMYLITA TALBOT [Born 1889; died June 01, 1983, Portola Valley, California] She was the mother of Charles Wilson, who worked for Pope & Talbot in both the steamship division and the lumber divison.

GEORGE ANDREW POPE SR. married EDITH TAYLOR April 26, 1892, in San Francisco, California, daughter of WILLIAM TAYLOR and MARY. She was born in 1868 in San Francisco, and died in December 1944 in San Francisco.

    EMILY TAYLOR POPE [Born December 03, 1898; died May 25, 1971, San Mateo, California] She was the mother of Emily Andrews, whose husband worked for many years for Pope & Talbot. Her last husband, Geoffrey Montgomery, was a director of Pope & Talbot.

    GEORGE ANDREW POPE JR. [Born November 12, 1901, San Francisco; died January 01, 1979] He was a longtime leader of Pope & Talbot. He consolidated his position by marrying his cousin Harriet Talbot Brownell.

GENERATION NO. 4

TALBOT CYRUS WALKER married MARY ALVORD KEENEY June 27, 1911, in San Francisco, California, daughter of JAMES KEENEY and MARY JONES. She was born in 1887 [location unknown], and died in 1965 [location unknown].

    CYRUS TALBOT WALKER [Born November 05, 1912; died August 12, 1999, Portland, Oregon] He worked for Pope & Talbot for many years, replacing his cousin George Pope as CEO.

SOPHIA GLEASON PIERCE married DR. EDWARD ERLE BROWNELL September 10, 1902, in San Francisco, California, son of WILLIAM BROWNELL and HARRIET LOWE. He was born March 26, 1874, and died May 03, 1939 in California.

    HARRIET TALBOT BROWNELL [Born September 06, 1907, San Francisco, California; died May 17, 1994, Palm Springs, California] She married her cousin George Pope. Her second husband, James Bamett, was a director of Pope & Talbot.

FREDERICK C. TALBOT was born in November 1895, in San Francisco, California, and died in 1955, in San Francisco, California. He married MADGE FISHER in 1918.

    FREDERICK C. TALBOT JR. [Born 1920, Aberdeen, Washington; died 1981, Redwood City, California] He worked for Pope & Talbot in the Northwest and in California in the lumber and steamship divisions.

EMILY TAYLOR POPE married MOSELEY TAYLOR 1914, son of WILLIAM TAYLOR and MARY MOSELEY. He was born in 1895, in Boston, Massachusetts, and died in February 25, 1951, in New York City.

    EMILY POPE TAYLOR [Born May 19, 1925, Boston, Massachusetts] She is a general partner of Pope Resources. She married Adolphus Andrews Jr., who was a director of Pope & Talbot as well as holding various management positions over many years. He also was a director of Pope Resources.

GEORGE ANDREW POPE JR. married HARRIET TALBOT BROWNELL September 06, 1930, in Trinity Episcopal Church, Gough & Bush Streets, San Francisco, California, daughter of EDWARD BROWNELL and SOPHIA PIERCE. She was born September 06, 1907, in San Francisco, California, and died May 17, 1994, in Palm Springs, California.

> **GEORGE ANDREW POPE III** [Born April 22, 1932] He was a director of Pope & Talbot, employed primarily in the forestry area.
> **PETER TALBOT POPE** [Born August 01, 1934] He was director of the company and succeeded Cyrus Walker as CEO. He is also a General Partner of Pope Resources and a director.
> **GUY BROWNELL POPE** [Born February 13, 1936] He was a director of the company as well as COO. He retired in 1978 due to illness.

GENERATION NO. 5

EMILY POPE TAYLOR married ADOLPHUS ANDREWS JR. June 28, 1947, in Burlingame, California, son of ADOLPHUS ANDREWS and BERNICE PLATTER. He was born March 22, 1922, in New York.

> **GORDON POPE ANDREWS** [Born January 07, 1957, San Francisco, California] He is a director of Pope & Talbot.
> **EDITH TAYLOR ANDREWS** [Born January 30, 1954, San Francisco, California] She married JOSEPH OLIVER TOBIN II April 08, 1989, in San Francisco, California. He is a director of Pope Resources.

PETER TALBOT POPE married JOSEPHINE DAY April 04, 1964, in Portland, Oregon. She was born December 06, 1936, in Portland, Oregon.

> **MARIA MACGREGOR POPE** [Born February 21, 1965 in Portland, Oregon] She is CFO of Pope & Talbot, Inc.

SELECTED BIBLIOGRAPHY

Barman, Jean. *The West beyond the West: A History of British Columbia.* Toronto: University of Toronto Press, 1991.

Bunting, Robert. *The Pacific Raincoast: Environment and Culture in an American Eden,* 1778–1900. Lawrence: University Press of Kansas, 1997.

Clark, Norman H. *Washington: A Bicentennial History.* New York: Norton, 1976.

Coman, Edwin T. Jr., and Helen M. Gibbs. *Time, Tide, and Timber: A Century of Pope & Talbot.* Stanford, Calif.: Stanford University Press, 1949.

Cox, Thomas R. *Mills and Markets: A History of the Pacific Coast Lumber Industry to 1900.* Seattle: University of Washington Press, 1974.

———. "Trade, Development, and Environmental Change: The Utilization of North America's Pacific Coast Forests to 1914 and Its Consequences" in *Global Deforestation and the Nineteenth-Century World Economy.* Richard P. Tucker and J. F. Richards, eds. Durham, N.C.: Duke University Press, 1983.

Cox, Thomas R., et al, eds. *This Well-Wooded Land: Americans and Their Forests from Colonial Times to the Present.* Lincoln: University of Nebraska Press, 1985.

Curtis, Robert O., and Andrew B. Carey. "Timber Supply in the Pacific Northwest." *Journal of Forestry* 94 (September 1996): 4–7, 35–38.

Eakins, Jan M. "Historic American Engineering Record: Port Gamble, Washington" HAER, August 1997.

Ficken, Robert E. *The Forested Land: A History of Lumbering in Western Washington.* Seattle: University of Washington Press, 1987.

Ficken, Robert E., and Charles R. LeWarne. *Washington: A Centennial History.* Seattle: University of Washington Press, 1988.

French, Michael. *U.S. Economic History since 1945.* Manchester, England: Manchester University Press, 1997.

Gibbs, Helen M. "Pope & Talbot's Tugboat Fleet." *Pacific Northwest Quarterly* 42 (October 1951): 302–324.

Guthrie, John A., and George R. Armstrong. *The Western Forest Industry: An Economic Outlook.* Baltimore, Md.: Johns Hopkins Press, 1961.

Hays, Samuel P. "From Conservation to Environment: Environmental Politics in the United States since World War II." In *Out of the Woods: Essays in Environmental History.* Char Miller and Hal Rothman, eds. Pittsburgh: University of Pittsburgh Press, 1997.

Labaree, Benjamin H., et al., eds. *America and the Sea: A Maritime History.* Mystic, Conn: Mystic Seaport, 1998.

Matthews, Frederick C. *American Merchant Ships, 1850–1900* Series 2. Salem, Mass.: Marine Research Society, 1931.

Morgan, Murray. *The Last Wilderness.* New York: Viking Press, 1956.

Patterson, James T. *Grand Expectations: The United States, 1945–1974.* New York: Oxford University Press, 1996.

Robbins, William G. "The Tarnished Dream: The Turbulent World of the Forest Products Industry in the Northwest." *Montana* 37 (winter 1987): 63–65.

Tillman, David A. *Forest Products: Advanced Technologies and Economic Analyses.* New York: Academic Press, 1985.

White, Richard. *Land Use, Environment, and Social Change: The Shaping of Island County, Washington.* Seattle: University of Washington Press, 1980.

———. "Poor Men on Poor Lands: The Back-to-the-Land Movement of the Early Twentieth Century—A Case Study" *Pacific Historical Review* 49 (February 1980): 105–131.

Williams, Michael. *Americans and Their Forests: A Historical Geography.* Cambridge: Cambridge University Press, 1989.

Williams, Richard L., et al. *The Loggers.* Alexandria, Va.: Time-Life Books, 1976.

Yonce, Federick J. "Lumbering and the Public Timberlands in Washington: The Era of Disposal." *Journal of Forest History* 22 (January 1978): 4–17.

INDEX

*Pages with illustrations are indicated by **boldface** page numbers.*

A

absorbent products, 131, 149–50, 151, 154
    disposable diapers, 131–32, 149–50, 151, 154
    tissues, 131–32, **133**, 149, 151, 154
Absoroka, **84**
Adams, William J., 8
Admiralty Hall, **36, 37**
Admiralty Tug Boat Company, 77
advertisements, **109, 147**
Alderwood Manor, 54, **72–73**, 75
Aluminum Company of America, 83, 85
American Federation of Labor, 56, 57
Ames, Edwin G., 33, 37, 38, 50, 53, 54, 55, 56, 57, 58–59, 70, 77
Ames, Maude Walker, 37
Andrew, Ralph W., 47
Andrews, Adolphus (Dolph), Jr., 88, **89**, 99, 105–6, 108, 110, 111, 132, **147**
Andrews, Emily, 88, **89**
Andrews, Emily T., 142
Atalanta, 40, 41

B

Balch, Lafayette, 5
Baldwin, Alex, **89**
band saws, 26, 27, 45, **96**
Big Mill, **66–67**. *See also* Saint Helens
Bonanza, 41
Bremerton, **87**
British Columbia, 110–12
Broadmoor, **58,** 59, 75
Brown, E. S., 6
Brownell, Bill, **89**
Brownell, Katy, **89**
Brownell, Sophia Talbot, **89**
buckers, **30,** 31
burners, **48**

C

Camp Cowlitz, 68–69
Camp Gamble, 68–69
Camp Talbot, 68–69, 75
Castlegar, 152–53
centennial, **89**
Central Lumber Company, 50
Chaloner family, 1
Charlesworth, Teddy, 76
Civil War, 17–19
Clark, Harry, **147**
clear-cutting, 122–23
Coman, Edwin T., Jr., 58
Condon, Richard, 59
consumer products, 149–51, 154
    disposable diapers, 131–32, 149–50, 151, 154
    tissues, 131–32, **133**, 149, 151, 154
Cooney, Neil, 47
Cosmopolis, 46–47
cranes, **60, 69**

D

disposable diapers, 131–32, 149–50, 151, 154
donkey engines, 30, 32–33
Douglas fir, clear-cutting, 122–23
Doumont, David, 126
Drew, Fred, 72
Dutton, Violet, **89**
Dwyer, James W., 122

E

East Machias, Maine, 1, **2**, 5, 15, 30
Eau Claire, **130,** 131–32, 150
Enoch Talbot, 77
environmental laws/restrictions/practices, 112, 145–46, 153–54, 157, 162–63
equipment: logging, 126, **152**
    lumber-making, 158
    mill, 25–26

F

faller-buncher, **125,** 126, **152**
Faught, Tom, **147**
felling, **22,** 31–33
Ficken, Robert E., 7
Fir-Text Insulating Board Company, 66
Fitzhenry, E. C., 39

Flannery, Michael, **147**, 151, 161
Folquet, George H., 110, 114, 121, 142
Foster, Charles, 5–6, 8, 40
Foster family, 1
Friesen, Abe, 162
Frohnmayer, Bill, 135

G

Gamble Bay, 6. *See also* Port Gamble
Gibbs, Helen M., 58, **89**
Gold Rush, 2, 7
Goliath, 76
Gove, William, 76
Grand Forks, 111, 117–18, **154**
Grays Harbor Commercial Company, **46**–47, 49, 51

H

Halsey, 129, 132, **134**–40, 154, 156–57, 160, 162
Hammersmith, W. N., 86
hardboard, **141**
Harmac Pacific, **160,** 162
Hauptman, Sidney M., 61, 63, 71, 74
Hayes, Roger, **147**
H. D. Bendixsen, **19**
headrigs, **98**
high-lead logging, 30, 33
Hirschi, Walter, 86
Hooper brothers, 38
Hudson, 118
Hunter, E. N. W. (Ed), 88, **89**, 103
Hunter, Mary, **89**
Hutton, Sophia, **89**

I

Indians, 6, 14, 42
Industrial Workers of the World, 55–57

J

Jackson, A. W., 38, 46–47
James Cheston, 41
Johnson, Clark A., 110

K

Kalama, 108, 141
Keller, A. W., 20
Keller family, 1
Keller, Josiah P., 5–6, 8, 13–14, 17–18, 21, 40
kilns, **65**
King Philip, 20
Kitsap, **20**

L

labor organizations, 44, 47, 55–57. *See also* strikes
Ladysmith, 131
land: acquisition, 19, 21–23; sales, 53–55
Lamadrid, Carlos, **147**
Libby, 76
"Little Boston," **28**, 42
log conversion, **108**
log ponds, **27**
loggers, **30, 31, 32,** 124–26
logging, **22, 93,** 152
    camps, 56
    clear-cutting, 122–23
    in the early twentieth century, 54
    equipment, 126, **152**
    high-lead, 30, 33
    in the nineteenth century, 30–33
logs
    converting to lumber, 158
    milling, 25–27
    towing, 76–77
    transporting, 32–33, **38,** 49, **52**
Lueddemann, Hillman, 74–75, 83, 85, 88, 106
lumber: carriers, 68, **69**; making, 158
Lumbermen's Protective Association, 56, 57

M

machine shop, **43**
Mackenzie, 162
Madison, Bob, **147**
Mahaffay, Robert, 86
Mason, R. Steven, **147**, 149
McCormick, Charles, 54, 59, 61, **63,** 64, 66, 68, 71, 74, 79
McCormick (Charles R.) Lumber Company, 63–64, 71, 74–75, 78, 79
    overlogging, 69–70

McCormick, Hamlin, 66
McCormick Intercoastal Steamship Company, 63–64
McCormick Shipping Company, 74, 79
McCormick Steamship Company, 61, 63–64, 71
    parade float, **62**
Medical Arts Building, **78**
medium density fiberboard, 99, 108
Michelson, Lewis (Dr.), **89**
Michelson, Vera, **89**
Midway, 111, 117–18, **159**
mills, **24, 25**, **43**, **51**, **95**, **118**, 152–55
    equipment, 25–27, 45
    modern, 158–**59**
    pulp, 135–**37**, **138**
    women working in, **84**, 85
    workers in, 86
    See also individual locations
Montgomery, Emily Pope, **89**
Montgomery, Jeff, **89**
Munson-McCormick, 61

N

Nanaimo, **160**
Newcastle, 151, 162
No. 2 Port Gamble, 25–27, 28. *See also* Port Gamble

O

Oakridge, 92, **95**–100, **109**, 117, 120, 122, 129, 133, 140–41
    last years, 145–48
Okanogan, **21**
Oliver, Egbert S., 47
Oregon, expansion into, 92–100
overlogging, 69–70

P

Pacific Argentine Brazil Line, 61–63, 74, 100–3
Pacific Atlantic Lumber Company, 74–75, 78
Pacific Coast Lumber Manufacturers Association, 55
Pacific Lumber Inspection Bureau, 55
Pacific Pine Lumber Company, 24, 44, 46, 49, 50
Pacific Pine Manufacturers Association, 24
paper
    making, 135, 138
    recycling, **130**, 132
particleboard, 96–97, 107

pellets, 148, **150**
Penn Tract, 92–94, 95, 113, 120, 143, 145, **147**
piles, **70**
Pioneer, 76
plywood, 107–**8**, 141
Point San Pedro, **81**
pollution control in pulp making, 135–37
Polly, 76
Polson, Alex, 64
pools, 24, 28
Pope, Andrew J., 2–10, 12, 13, 15, 17, 29, 40
Pope & Talbot (Inc.)
    adoption of new logging technologies, 124–26
    advertisements, **109**, **147**
    attempted sale of, 53–54, 57–58
    attempted sale to ALCOA, 83, 85
    centennial, **89**
    change to western ownership, 17–19
    company timeline, ix-xiii
    consumer products, 149–51, 154
    core sectors, 140–41
    departure from San Francisco, 106
    effect of World War II on, 83–85, 92
    end of shipping operation, 100–3
    environmental laws' effect on, 112
    establishes Port Gamble site, 5–7
    expansion into British Columbia, 110–12
    expansion into Oregon, 92–100
    founding and early days, 1–15
    growth, 35–59
    incorporation, 79–80
    increased revenue from manufactured wood products, **107**–8
    land acquisitions, 19, 21–23
    land development, 59
    land sales, 53–55
    move to full utilization, 96–100
    perception as shipping company, 39
    post–World War II holdings, **92**
    post–World War II period, 88–103
    product diversification, 127–32
    property development and home building, 112–13, 114
    purchase of public timber, 120, 121, 127
    recession's effect on, 132–34
    regained by original owners, 74–75, 78–80
    reincorporation, 128
    renamed W. C. Talbot & Company, 6

INDEX **171**

reorganization, 49
repositioning, 145–60
restructuring, 151
sale of, 59
second generation takes over, 17–29
ships, 40–41, 45, 77
spin-off of Pope Resources, Inc., 142–43
under McCormick Lumber Company, 61–64, 67, 68–74
Pope & Talbot Development, Inc., 110, 113
Pope & Talbot Lumber Company, 67
  incorporation, 79
Pope, Edith, 105
Pope, Emily Talbot, 8, 29
Pope family, 1, 54; descendants, 164–67
Pope, George A., Jr., 80, 83, 88, **89**, 102–3, 105–7, 117
Pope, George A., Sr., 29, 36, 49, 53, 54, **59**, 64, 71, 74, 80, 83, 85
Pope, George III, 106
Pope, Guy, 100, 106, 107, 108, 110, 117, 118, 127, 131, 132, **147**
Pope, Harriet Brownell, 54, 80
Pope, Kenneth, 80, 105
Pope, Maria M., 161, 162
Pope, Peter T., 88, 106, 107, 112, 117, 120, 127, 131–32, 142–**43**, 146–**47**, 149, 151, 161
Pope Resources, Inc., 110
  spun off from Pope & Talbot, Inc., 142–43
Pope, Samuel, 18
Pope, William, 18
Pope, Zib, **89**
Port Gamble, 6–8, **11–15**, 17, **29**, 39, **42, 43**, 50, 52–53, 58, 63, 64, 68, **69**, 75, 83, 86, 97, 99–100, 117, **118**–20, 140–41, 143; closure, 148, **149**; No. 2, 25–27, 28
Port Ludlow, 26, 27, 28, 39, **42, 48**, 49, 50, **51**, 52–53, 58, 64, 68, 75, 113, 114–**15**, 133, 143
Primrose, George, 76
public timber, 120, 121, 127
Puget Commercial Company, 49
Puget Lumber Company, 49
Puget Mercantile Company, 49
Puget Mill Company, 5–9, 11, 14, 18–19, 44, 46, 49, 53, 54, 63–64, 70–71
  acquisition of timberland, 19, 21–23
  change to western ownership, 18–19
  early-twentieth-century logging practices, 54
  incorporation of, 29
  land development, 72–73
  membership in pools, 24, 28
  mills, 25–27
  nineteenth-century logging practices, 30–33
  railroad building, 33
  tow and tugboats, 76–77
Puget Shipping Company, 49
Puget Sound Commercial Company, incorporation of, 29
Puget Sound Towage Company, 49, 77
Puget Sound Tug Boat Company, 49, 77
Puget Trading Company, 49
pulp
  cycles, **129**, 131, 157
  making, 128–31, **134**–40, **156–57**, 160, 162
  mills, 135–**37**

R

railroads, 33, 42–44, 69. *See also* trains
Rainier, 41
Rainier Investment Company, 49, 63–64, 71
Ransom, 149, 150, 151
recycling paper, **130**, 132
reforestation, **123**
Resolute, 76
Rolling Green Estates, **113**

S

Saint Helens, 63, 64, **65, 66–67**, 68, 70, 74, 75, 83, 85, 99–100, 108, 117
Saint Helens Pulp and Paper Company, 66
Sanborn, Lucius, 3
San Francisco, **3, 4, 18**, 39; earthquake, 51–52
sawmills. *See* mills
saws, 25–27, 45, **75, 93, 96**, 45
ships, **160**
  Absoroka, **84**
  Atalanta, 40, 41
  Bonanza, 41
  bought by McCormick Shipping Company, 79
  dangers facing, 40–41
  end of fleet, 100–3
  end of Pope & Talbot fleet, 100–3
  Enoch Talbot, 77
  Goliath, 76
  H. D. Bendixsen, **19**
  James Cheston, 41

King Philip, 20
Kitsap, **20**
McCormick, 61, **62**
McCormick, **71**
Okanogan, **21**
Pioneer, 76
Point San Pedro, **81**
Polly, 76
post–World War II routes, **101**
Rainier, 41
Resolute, 76
steam schooners, 81
Shore Woods, 113
skid roads, 32
spar skidders, **65**, 69
spar trees, **31,** 33
Spearfish, 140, 151, 162
steam schooners, **81**
Stewart, Loran "Stub," 92, **147**
Stout, Jim, **147**
strikes, 100–2, 151. *See also* unions
Sutherland, **146**

T

Talbot, Charles F. A., 38, 39, 50–51
Talbot family, 1; descendants, 164–67
Talbot, Frederic, 2–4
Talbot, Frederic C., 29, 38, 41, 47, 49, 80, **89**
Talbot, John, 40
Talbot, Ted, **89**
Talbot, Vera, **89**
Talbot (W. C.) & Company, 6
Talbot, William C. (Captain) , 4–10, 13, 17, 29, 41
Talbot, William H., **28**, 29, 35, 36, 37, 38, 41, 44–45, 46, 49, 50, 53, 54, 71, 73, 74
Teekalet Bay, 5–6, 21. *See also* Port Gamble
Thompson, W. A., 86
Thompson, Will, 47
Thrasher, Frederick (Captain), 40, 41
timberland, acquisition of, 19, 21–23
timeline, ix–xiii
tissues, 131–32, **133**, 149, 151, 154
towing logs, 76–77
tractor, **52**
trains, 32, **34**, **38**. *See also* railroads
tree farms, 95, **120,** 121, 122
trees: felling, **22**, 31–33; spar, **31,** 33
tug boats, 76–**77**, **127**, **146**

U

unions, 44, 47, 55–57, 79, 146. *See also* strikes
Upper Willamette Tree Farm, 95
Utsalady, **24**, 26, 28, 49, 50

V

Van Syckle, Edwin, 46
veneer, 96–99, 107

W

Walker, Cyrus, 6, **9**, 10, 12, 13–14, 18–19, 23, 26, 29, 35–36, 37, 38, 41, 44–45, 49, 50, 53, 54, 77, **89**
Walker, Cyrus T., 95, 105, 106, 107, 112, 117, **147**
Walker (Cyrus T.) Tree Nursery and Forest Research Center, **119, 122**
Walker, Talbot, 80, 83
Walker, Talbot C., 29, **89**
Walker, William, 26, 35, 37
Washington Park, 53–54, 72
W. C. Talbot & Company, 6
West Coast Lumbermen's Association, 55
Weyerhaeuser, 50–51, 64
Wheeler, Charles L., 75, 83, 85, **89**, 90–91
Wheeler, Mrs., **89**
Whelan, William A., **128**, 140, 149
White Gold 49, 138, **139**
White, James, 6
Wilburn, John W., 40
Wilson, Amylita Talbot, **89**
Wilson, Charlie, **89**
Wilson, Winnie, **89**
Wobblies, 55–57
women, working in mills, **84**, 85
wood pellets, 148, **150**

Y

yarding, 32–33, **33**